WORLDWIDE TOP BOUTIQUES

深圳市创扬文化传播有限公司　编

国际顶级服饰名店

海峡出版发行集团 | 福建科学技术出版社
THE STRAITS PUBLISHING & DISTRIBUTING GROUP | FUJIAN SCIENCE & TECHNOLOGY PUBLISHING HOUSE

前言 FOREWORD

服饰专卖店是服饰营销渠道的重要组成部分，是目前服饰零售中比较有效的一种方式，它是将服饰商品传递给最终消费者的最直接场所，也是生产者快速获取市场真实信息的重要渠道。

许多服饰厂商将专卖店作为品牌建设的基点，进而完善整个大品牌体系，以设计视角综合品牌的风格定位、视觉识别（标志、色彩）、经营理念等作为品牌要素，并把这些要素以最优化手法渗透到专卖空间设计中。

服饰专卖店的设计成功与否，不仅影响到品牌的现实利益，而且也关系到品牌的发展和延伸。服饰专卖店设计中服饰品牌定位、设计品位，以及延伸出来的流行特征都直接影响到空间形象，成功的空间设计应该做到能营造出品牌定位、设计品位和品牌背后演绎给消费者的生活概念和文化理念。店堂的店面设计要完善并能体现品牌的风格，衬托出品牌主题，并能很好地进行融合，这直接影响到品牌经营的成败。在空间设计中功能性的作用直接对空间的服务带来影响，合理的空间布局和结构设计将会对销售活动起来促进作用。合理地选择装饰材料，运用施工工艺，不仅为设计构思的实现提供可行性，而且也将成为整个品牌形象的一个重要组成部分。服饰专卖店设计中经营理念和设计理念的结合直接影响到空间定位，准确的市场定位将会使品牌的投入实现市场的最大效益。

本书精选了39个全球著名设计师最新设计的服饰专卖店案例。书中这些案例均为美国、欧洲、南美洲、亚洲等地最具有实力和最富创意的专卖店设计师的杰作。这些案例中，空间如何被有效利用，怎样体现空间的品质及树立高水准的视觉效果，如何营造出彰显服饰品牌风格和个性，设计师都进行了整体斟酌，使之成为服饰专卖店设计的典范。

本书旨在为读者提供一个风格各异、特点鲜明的服饰专卖店设计案例，欣赏性及参考性兼具。

编者

2011年4月

Clothing fashion boutique is an important component of clothing marketing channel, one of the effective clothing retail modes at present. It is not only the most direct place to display the clothing commodities to ultimate customers but also an important channel for producers to get access to real market information rapidly.

Many clothing manufacturers take fashion boutiques as foundation of brand building to perfect big brand system. From perspective of design, the combinations of style positioning of brand, visual identity (logo and color) and operation philosophy etc acts as brand elements which permeate into the space design of fashion boutiques by optimization method.

Whether the design of clothing fashion boutique is successful or not will not only influence the realistic benefit of a brand, but concern the development and extension of a brand as well. In the design of clothing fashion boutique, the clothing brand positioning, design style and the outstretched fashionable features have direct influence on space image. Successful space design should succeed in creating the life concept and cultural idea that the brand positioning, design style and brand express to the customer. The design of storefront should be impeccable, showing the brand style and highlighting the brand theme while merging them together in a good way, which has an immediate impact on the brand operation. Functionality in space design has a direct effect on space service. Reasonable space planning and structural design will promote sales. Reasonable choice of decorative materials and application of construction technology can make it feasible to realize the design concept and will be an important component of the entire brand image. The combination of operation philosophy and design concept in design of clothing fashion boutique has a direct impact on space positioning. Accurate market positioning can make the brand investment realize benefit maximization in market.

The latest cases of clothing fashion boutiques designed by 39 world famous designers were selected elaborately in this book. The cases in this book are all masterpieces of most talented and creative boutique designers from the USA, Europe, South America and Asia etc. In these cases, the designers made thorough considerations on how to effectively use the space, how to show the character of space and create high-standard visual effect and how to highlight the style and character of brand, which made these cases models of clothing fashion boutique design.

This book is aimed at providing readers with design cases of clothing fashion boutiques with various styles and distinguished characters, which are both appreciative and referential.

Editor

April 2011

目录 CONTENTS

Hermès Rive Gauche

设 计 师：Denis Montel, Dominique Hébrard, Sybil Debu, Mathieu Alfandary
设计公司：RADI
项目地点：Paris, France
建筑面积：2155m²
摄　　影：Bruno Clergue, Michel Denance

Hermès商店的原址是一座游泳池。巨大的空间十分宽敞，给人一种比真实空间更大的感觉。
这个项目设计有两个目的。首先是尊重、保留和重新解释这座游泳池的建筑。唯一重要的改变是通过轻结构支撑的混凝土水泥板将游泳池覆盖。下方则完整保留了整个游泳池。另一方面完全是现代的。这个部分由四个庞大的白蜡木屋构成，它们打破了原有的内部构造，并且与之形成了反差。原本的游泳池被这些灵活、轻便而且可活动的木屋取代，显示出了房屋内部的创造性。
入口就像倒置的采光井，与地面平行的设计吸引人们径直走向内部的光亮处，走向原本游泳池的地方。为了引导顾客，突出了入口景观并且做了极小的收缩，十分像卢森堡花园里的Médicis喷泉的四面。天花板微微倾斜，墙壁呈曲线形向内倾斜，覆盖着白蜡木板条，使凹处敞开如同漂浮着似的。神秘的色彩吸引着人们进入这家新店。

Hermès is setting up shop in a swimming pool... An immense volume, empty. An impression more of space than of surface area.
The project has a double aim. First of all to respect, conserve and reinterpret the architecture of the swimming pool. The only important modification was the covering of the pool by means of concrete composite floor slab supported by a light structure. Underneath, the pool has been integrally preserved. The facade, giving onto the rue de Sèvres, has kept its original appearance. Then, to tell another story, one that is resolutely contemporary. This takes form through the appearance of three monumental ash huts which both disrupt the existing volumes and converse with them. The invasion of what was once the pool by these huts, flexible, light and nomadic, suggests the creation of houses within the house.
The entrance is like a lightwell overturned, horizontal, which attracts one irrevocably towards the light at the back, towards what was the Lutétia swimming pool. To guide the visitor, the perspectives are accentuated and modified by an imperceptible contraction, rather like the sides of the Médicis fountain in the Luxembourg garden. The lightly inclined ceiling, the walls curved and leaning inwards, covered with oak laths that leave recesses open as if floating in matter. An introduction full of mysteries inciting one to plunge into this new Hermès house...

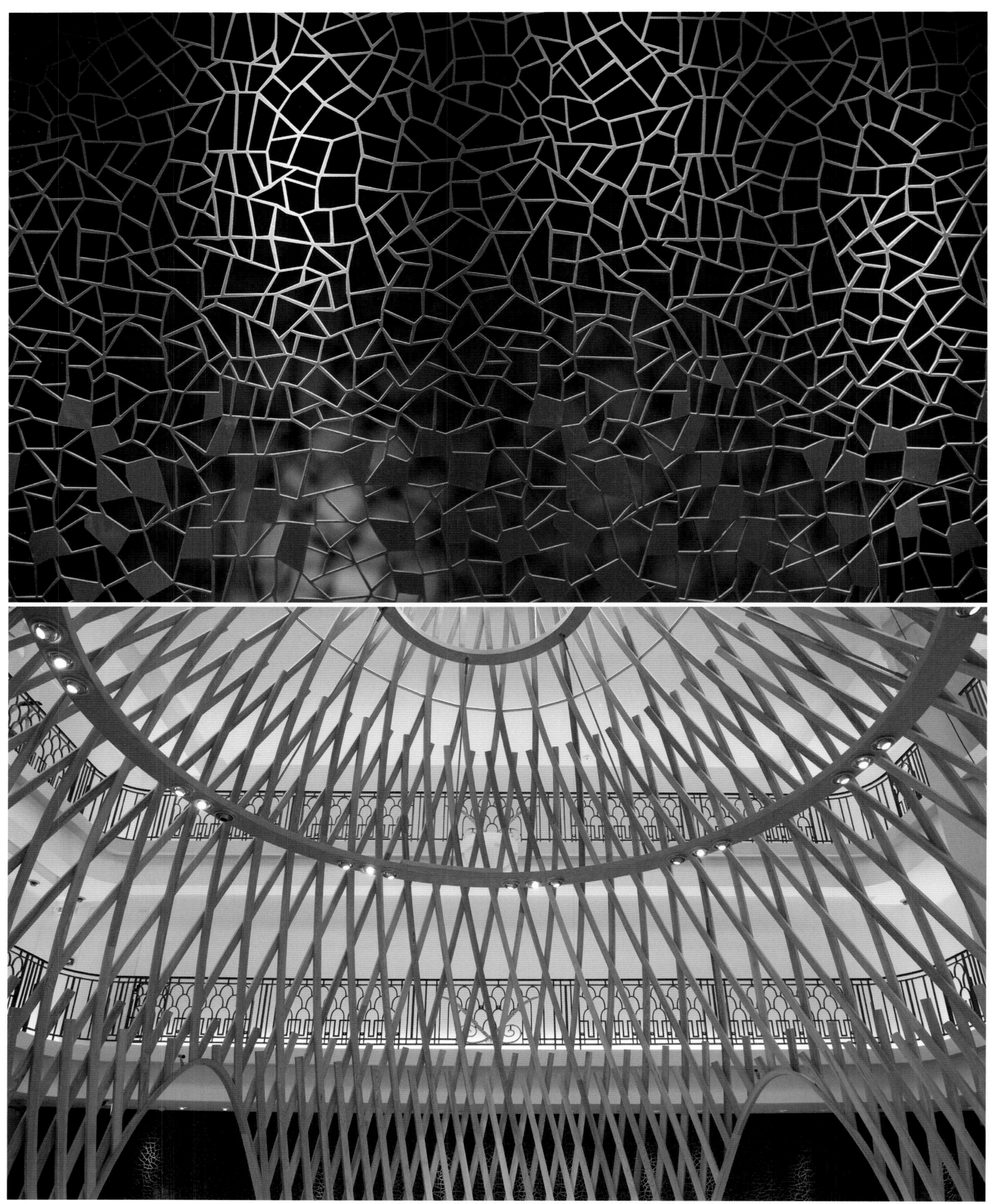

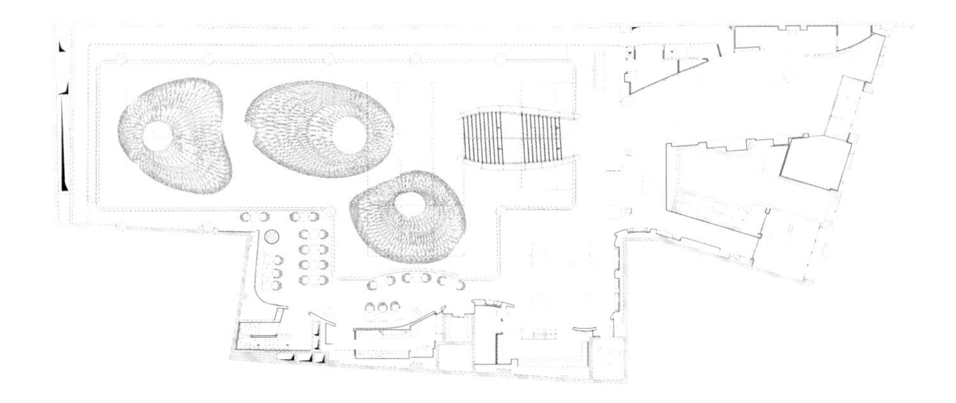

有些人会把这些占据了游泳池空间的小屋比作鸟的巢穴。这些不同形状、不同大小的小屋是用白蜡木材做成的。它们是搭建在一个由两个圆曲线构成的、木条网编织的自支撑结构里。

小屋里存放着Hermès的商品系列。他们似乎轻易地飞落在地上，增添了这个项目的灵活空间。第四个小屋看上去似乎是躺倒着的，与电梯形成一线，很自然地带领着顾客们朝着游泳池的方向走去，在入口和游泳池的空地处形成了连接。

Others will liken these huts, which occupy the volume of the swimming pool, to the nests of tisserin birds. These pavilions of different form and dimensions are constructed in ash wood. They are self-supporting structures that rest on a system of woven wooden laths with a double radius of curves.

he huts house the Hermès collections. They seem to have simply alighted on the ground, lending the project its nomadic dimension. The fourth hut, which appears to be lying down, lines the staircase that naturally leads the visitor towards the pool and forms the link between the entrance and the open space of the swimming pool.

这样的空间，灯光十分关键。整个房间沐浴在自然光中，光线是从中庭上的三个大天窗透进来的，只通过金属屏使其柔化。到了晚上，天窗都会被点亮，以避免产生“黑洞”的效果。
蔓延整个底层的白色石膏粉刷的波浪起伏的墙壁，由LED灯条从上方照亮，而光源则被隐藏起来了。小屋从内部点亮，看上去就像是巨大的灯笼。镶嵌在地板上的照明装置，照亮了小屋的网格状木质拱顶。每一个小屋都有一个由双环状悬浮木组成的大型的枝形吊灯。架子由集成的隐形LED灯条照亮。

In such a volume, the lighting is crucial. The entire space is bathed in natural light that penetrates through the three large skylights above the atrium, softened only by a metal screen. At night the skylights are lit to avoid a "black hole" effect. In order to avoid putting the spaces overlooking the pool that previously housed the changing rooms, in the shade, the effects had to be measured out, the contrasts that would otherwise have been too harsh attenuated. All the vertical panels are therefore also lightly illuminated.
The undulating walls in white plaster, running around the ground floor, are lit from above by LED tape with the light source hidden from view. Lit from the interior, the huts appear as giant lanterns. A lighting device embedded in the floor, illuminates their vaults of latticed wood. Each hut has a large chandelier composed of a double ring of suspended wood. The shelving is lit by integrated and invisible LED tape.

Tally Weijl

设计公司：Dan Pearlman
公司网站：www.danpearlman.com
项目地点：Paris, France

充满迪斯科球、聚光灯、闪光、镜子的世界，年轻的时尚品牌Tally Weijl邀请他的自信的主角们踏入这个反映他们生活方式的世界。Tally Weijl让那些无忧无虑的时尚追随者成为这个舞台的明星， 共同庆祝他们所定位的青少年市场的玩乐主义。

这个新兴的闪闪发光的Tally Weijl世界，是基于Dan Pearlman的设计理念在柏林、伯恩、巴塞尔形成的。在巴黎这个创意中心重新塑造了Tally Weijl的品牌形象，然后将这一新的形象传递给所有新开张的Tally Weijl店面。

外观轻盈的设计传递出了商店的规模，在人们进入这个空间之前就留下了深刻的印象。通过照明广告的应用将关注的焦点引向收银台附近，亮色的墙壁使服装的陈列更加突出显著，这些都使得所有元素之间产生了令人影响深刻的戏剧性效果。它引导着顾客在商店里穿行，而且很容易让顾客在商店各种不同的区域发现新东西。

With disco balls, spotlights, glitter, and mirrors, the young-fashion brand Tally Weijl invites its self-assured protagonist to step into a world that reflects their lifestyle. Tally Weijl makes carefree fashionist the stars of the show and celebrates the "fun-loving" credo of the target teenage-market segment. Of course-as distinctive brand elements-the bunnies shouldn't go without mention here.

The new glittering worlds of Tally Weijl came into being in Berlin, Bonn, and Basel based on a design concept envisioned by Dan Pearlman. The approach was to refashion the outlook of the brand in its creative epicenter of Paris, and then to transport this vision to all new Tally Weijl shops.

With its airily designed facade that conveys a sense of the store's scale, a strong impression is made before one even enters the space. Via the use of illuminated adverts to focus attention around the checkout counters, and brightly painted walls for presenting clothing in prominent ways, a highly effective, theatrical interrelationship of elements emerges. It guides the customer around the store and allows for the discovery of new things to the left or right in various sections of the store.

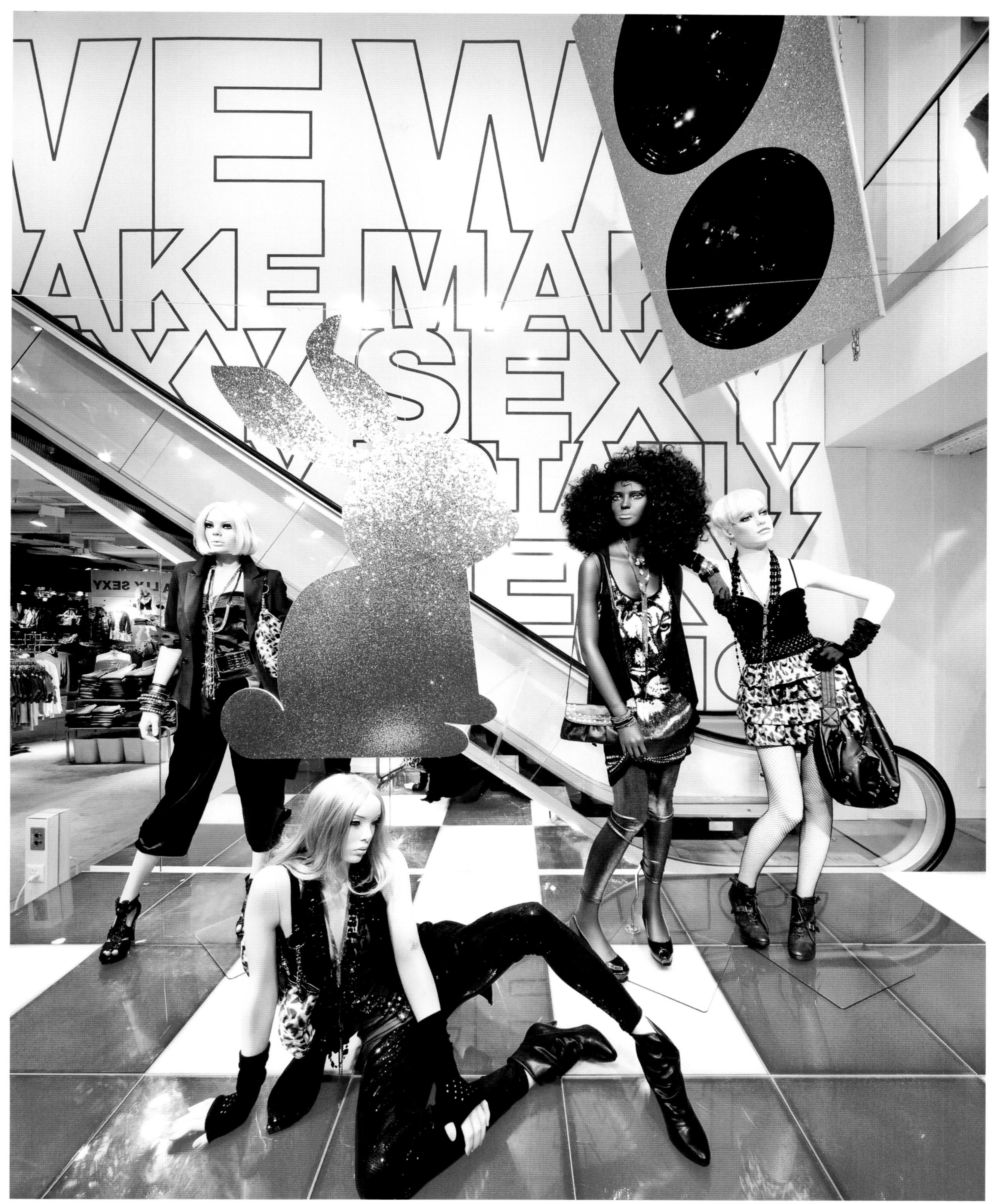
SEXY

从底层到一层的楼梯加宽了，并且加上了新的装饰，开辟出了突出区域。通过这样的扩张和亮色的使用，塑造了一个“舞台”，在建筑方面表达出了一种生机勃勃的音乐世界和俱乐部文化氛围。俱乐部风格的设计还包括了铬灯箱，超大的闪闪发光的喇叭以及自动扶梯和电梯区那些海报体的印字，这些构成了完整的设计风格。

相反的，休息区和更衣室能够让顾客舒适地徘徊其间，放松地试穿衣服。舒适的切斯特菲尔德风格坐椅为顾客提供了休息的地方，而休息室提供了更加私密的氛围。休息区模仿法国城堡的少女闺房设计，甚至能让经过了一整天购物疲累的人得到彻底放松。

The stairs leading from the ground floor to the first level were widened and new, additional decoration and highlight zones were created. Via this expansion, as well as the use of bright colors, a "stage" was fashioned that expresses in architectural terms the musical world and club-culture vibe of the exuberant, self-assured clientele.

The club design also includes chrome light boxes and oversized loudspeakers, which-covered in glitter and in combination with poster-like lettering in the escalator and elevator zones-contribute to an authentic, overall style.

The lounge-/chill out- and changing-room area, by contrast, invites customers to linger comfortably and to try on clothes at ease. Comfy Chesterfield-style seating offers a spot to catch one's breath, while the changing rooms offer a more intimate atmosphere. Modeled after boudoirs from an eighteenth-century French castle, the spaces allow even the most stressful day of shopping to be enjoyed to the fullest.

OTALLY
TALLY WEiJL
IS IT HOT IN HERE
OR IS IT ME ?
lift

ACCESSORIES

ACCESSORIES

ACCESSORIES
MAKE SEXY TOTALLY
IS IT HOT IN HERE

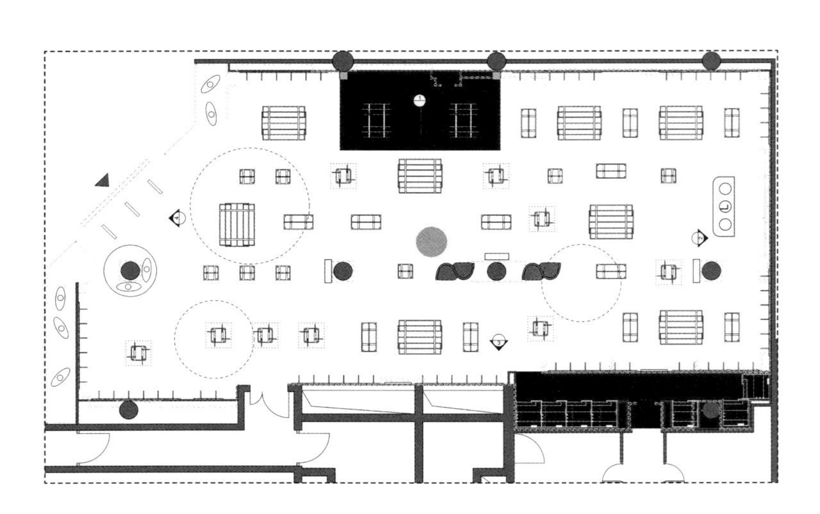

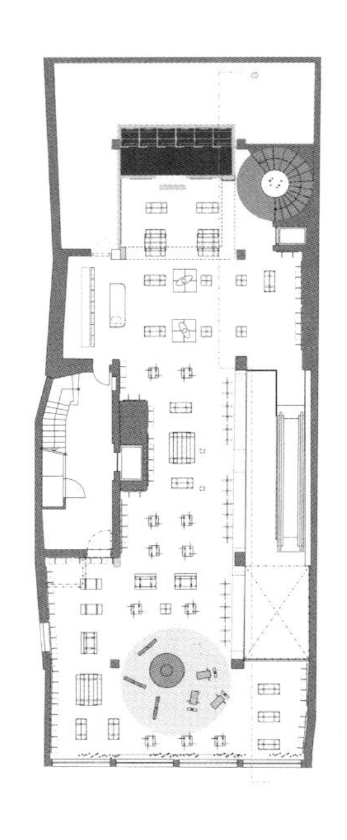

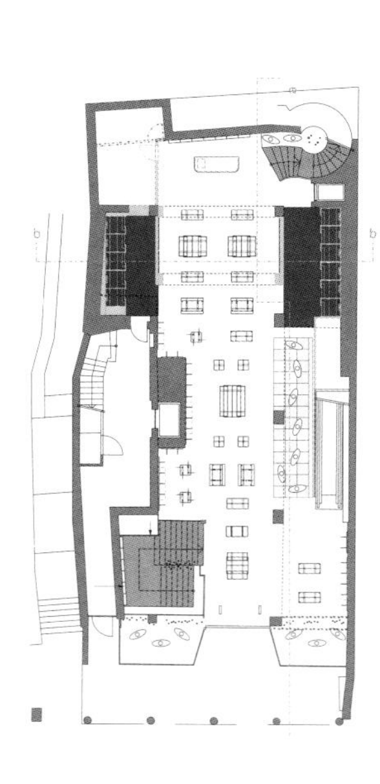

Colin's

设计公司：Dan Pearlman
公司网址：www.danpearlman.com
项目地点：Istanbul, Turkey
建筑面积：450m²

Colin's是一个土耳其的牛仔品牌，在很多国际市场都能看到。在最短的时间间隔里，Colin's成了俄罗斯排名首位的牛仔品牌。

这个品牌现在需要一个商店系统来支持这种动态的扩张，同时在牛仔领域传递一种休闲的生活方式及专业知识。

设计师以一种简单、大型的钢柜系统作为主要元素，展示了牛仔领域的多样性及优势。这种美学是受码头及产业场所的启发，并集中于一些主要区域，如街道、住宅和产业。这种概念最先在450平方米的旗舰店实践，然后再实施到各个连锁商店。

Colin's不仅仅是个牛仔品牌，它代表了一种生活方式、品质和设计。目标是将Colin's的精神和品牌价值转化成一种与众不同的可视语言。

将品牌下面的独立个体打造出一种高识别价值是非常重要的。品牌年轻而富有朝气的形象体现在连锁店的设计上。

Colin's is a Turkish jeans brand that is successfully represented in numerous international markets. Within the shortest space of time, Colin's established itself as the number 1 jeans brand in Russia.

The brand now needs a store system that supports the dynamic expansion and at the same time communicates a casual lifestyle and real expertise in denim.

Dan Pearlman introduced a simple, massive steel shelf system as a key element. This demonstrates variety and strength in the jeans sector. The aesthetics are inspired by docks and industrial sites and focus on the main areas of "street", "home"and "industrial"

Initially this concept should be realized on the example of a 450m² flagship store. Afterwards the store concept should be transferable to other store formats and shop in shop systems.

Colin's is more than a jeans brand: it represents lifestyle, quality and design. The aim is to transfer Colin's spirit and brand values into a distinctive visual language.

It is important to constitute the brand individual identity with a high recognition value. The young, dynamic, look of the brand should be reflected in the design of the stores.

COLIN'S

LOVE
32 USA

FRAGILE
HANDLE
WITH CARE

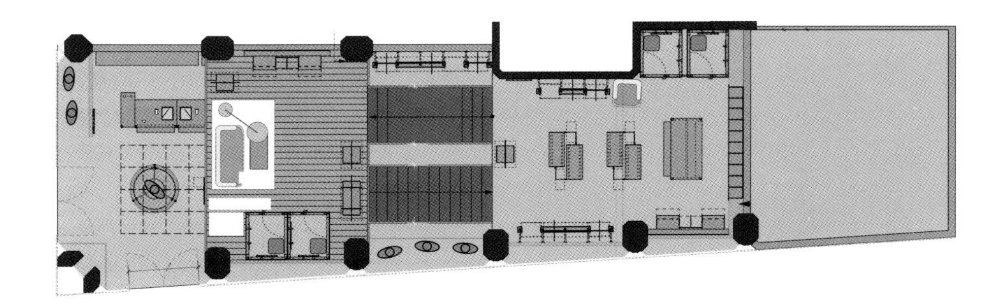

COLINS
COLIN'S
THE WORLD
COLIN'S
URBAN
COLIN'S
COLIN'S

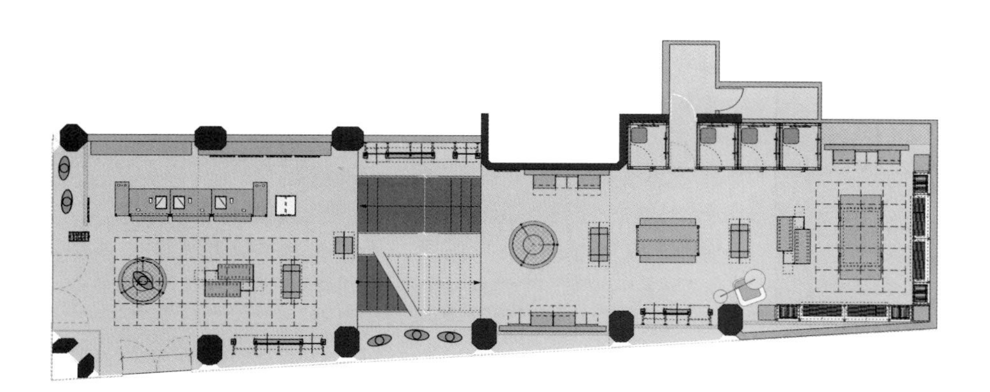

Holt Renfrew

设 计 师：Diego Burdi, Paul Filek, etc
设计公司：Burdifilek
项目地点：Toronto, Canada
建筑面积：Designer Collection Flooy 1500m²,
Personal Shopping 303m²
摄　　影：Ben Rahn, A Frame

Burdifilek的室内设计理念是将Holt Renfrew的豪华时装系列以一种纯洁而复合的白色调展示出来。这一设计理念重点在于追求奢华购物感受，将艺术感和戏剧感融入时装零售。白雪、珍珠和雪花石膏的色调以雕塑的形式和结构材料展示出来。4000根白色棍棒悬挂在圆形天花板凹槽里，微微地摇晃着，创造出了一个飘渺的背景。

整个气氛被设计成一个画廊的风格。在那儿，标签的选择能够展示出他们的优雅和丰富。悬挂的雕塑装置唤起了极强的阴柔感，形成了记忆点也激发了想象。圆形的时装系列区域划分和空间创造了贯穿整个楼层的许多记忆点。衣服展示在磨砂钢落地卡具上，并且以和天花上的细节一致的形式悬挂。

一幅透明喷砂的树脂立方体装饰的雕刻屏风从天花上通路而下，充满了诗意。所有的装置都笼罩在白光和散光照明里，创造出了空间的艺术感和飘渺感。各个不同的特色展桌发挥了展示产品的作用，它们本身又都是艺术品。大胆的材料应用，例如坚实的橡木伸出一只经过18K金刷面处理的手臂，染成紫红色的泰柚木和瓷器打磨的细节墙饰，造就了一个极度精致的建筑表情。

Burdifilek "interior concept displays Holt Renfrew" collection of luxury clothing labels in a pure and complex palette of whites. The design concept focuses on evolving the luxury shopping experience and infusing a sense of art and drama into fashion retailing. Tones of snow, pearl and alabaster play out in sculptural forms and textural materials. 40000 pristine white rods hang in suspended animation in monumental circular ceiling coves and sway gently, creating an ethereal backdrop.

This atmosphere is designed as a gallery where a selection of labels can present themselves in refined opulence. The hanging sculptural installation evokes ultra feminine volumes, creates a point of memory and inspires the imagination. The circular forms compartmentalize the collections and the space creating points of memory throughout the floor. Clothing is displayed on brushed steel floor-mounted fixtures and hang in unison with the ceiling details. This space not only looks, but feels a million miles away from outside world, and emphasizes a richly indulgent experience.

A sculptural screen of clear and sandblasted Lucite cubes fall poetically from openings in the ceiling details. The installations are bathed in white light and the diffused illumination creates an artistic and ethereal perception of the space. Throughout the floor, different feature tables function as display for product, but act as art pieces as well. Indulgent materials like solid oak with a hand applied 18 Karat gold dry brushed finish, aubergine stained Tay wood and porcelain finished wall details all contribute to a the highly detailed architectural expression.

DONNA KARAN

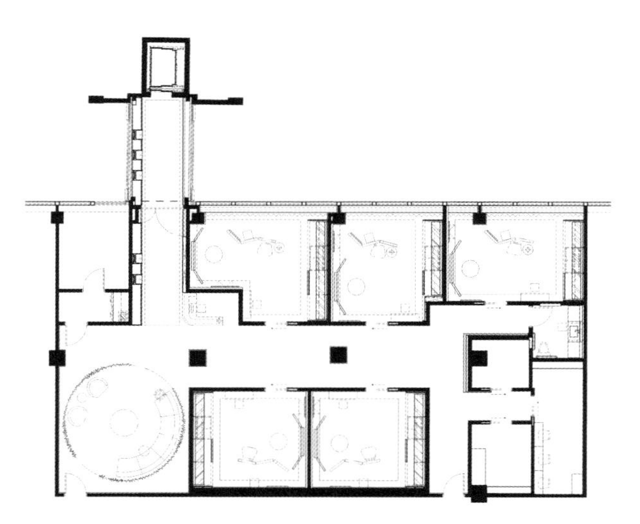

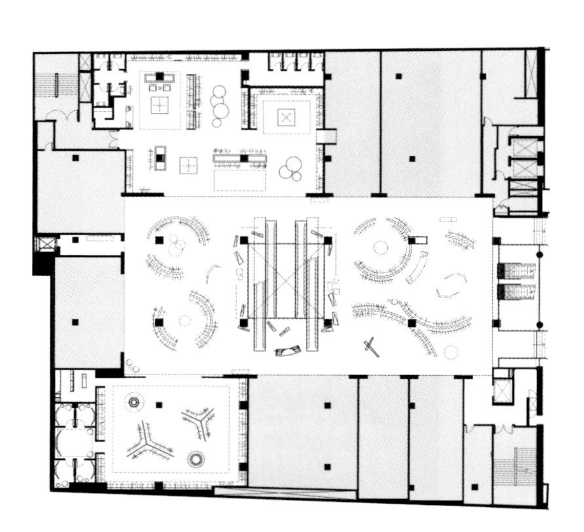

在这个设计中，接待处的旁边有五间私人套房。顾客们从玄关进入这个空间，一侧是惹人注目的摆着流线型几何图形的钢制雕塑的景观墙。这个雕塑背面打着光，前面是一架乳白色的玻璃飞机。在接待处那一边每一间套房都配备高贵的用品：定制的羊毛地毯，漆器屏风，从地板到天花的三面镜子，以及具有现代美感的定制家具。服装展示区就像一个小剧院，在这里造型师们可以为顾客一套一套地搭配出各种不同的造型。

Within the concept, there are five private suites located adjacent to a reception area. Clients enter the area down a hallway, flanked on one side by a dramatic feature wall with a streamlined geometric steel sculpture which is backlit behind a plane of opalescent glass. The reception area is a dramatically lit circular area contoured by thousands of stainless steel rods hanging from the ceiling detail to just inches above the floor. A custom sofa which is upholstered in a latte-coloured mohair fabric is perfectly contoured to the hanging sculptural installation. Beyond the reception area, each suite is outfitted in noble finishes like custom wool carpets, lacquer screens and floor to ceiling tri-mirrors, and are outfitted with custom furniture with a modern and indulgent aesthetic. The clothing display area is a "ini theatre" where stylists can assemble different looks for the customer to see them one at a time.

Brown Thomas Luxury Hall

设 计 师：Diego Burdi, Paul Filek, etc
设计公司：Burdifilek
项目地点：Dublin, Ireland
建筑面积：539m²
摄　　影：Ben Rahn, A Frame

这一特色空间参考了Brown Thomas品牌精致的现代主义风格，并且捕捉到了美丽的Brown Thomas产品的精髓。提升奢华的购物体验以及融入微妙精致的优雅感是设计概念的核心。

豪华大厅设计成一个中央广场，主要区域之外为Cartier和Tiffany & Co.等品牌的展示区域。四周环绕着双色玻璃，颜色会随着光线和顾客的经过而发生微妙的变化。隐形的悬浮玻璃橱窗吊在调整过的透明玻璃墙上，这样通过玻璃橱窗就可以看见毗邻的店中店。地面铺设着斑驳的灰褐色、奶油色和咖啡牛奶色的亚光大理石，柔光灯更突出了整个空间的金色彩虹效果。

半圆形的无缝玻璃展示柜似乎都漂浮了起来，悬挂在喷砂面的香槟色玻璃基底上。定制的牡蛎色山羊皮展示台为珠宝提供了一个低调的背景，使珠宝更加引人注目。优雅的曲线能使人们漫步观赏，轻易地吸引顾客进入这个空间。

Bertoia风格的雕塑装置作品从地板延伸到天花，给整个空间内部注入了动能。手工铰接着镍杠似乎被优雅地定格在暂停的动画中，闪耀着光芒。

在都柏林的旗舰店里有着与众不同的风格，它代表着Brown Thomas不懈地致力于高端零售业。

The signature space references the sophisticated modernism of the Brown Thomas brand and captures the essence of the beautiful products it houses. The design concept focuses on elevating the luxury shopping experience and instilling within it a sense of subtle, nuanced elegance.

The Luxury Hall is designed as a central piazza with enclaves for brands like Cartier and Tiffany & Co. projecting beyond the main area. The perimeter is sheathed in subtly reflective dichroic glass; the colour delicately shifting with light and movement as shoppers move through the space. Invisible floating vitrines are suspended on modulated transparent glass walls, allowing for a glimpse through to the adjacent shop-in-shops. Underfoot, honed marble flooring in mottled creamy shades of taupe, cream and café-au-lait was laid, while soft lighting further accentuates the shimmering gold-hued iridescence throughout.

Semi-circular seamless glass display cases appear to float, cantilevered over sandblasted champagne-coloured Starfire glass bases. Custom oyster-coloured suede displays provide an inconspicuous backdrop for the jewellery itself, allowing the product to take center stage. The graceful curves encourage meandering exploration and effortlessly lead shoppers through the space.

Infusing the interior volume with a kinetic energy, Bertoia-inspired sculptural installations extend floor to ceiling. Hand-articulated in polished nickel the rods appear to be elegantly captured in suspended animation, glinting off one another.

The Luxury Hall has a distinctive voice within the Dublin flagship, and reflects Brown Thomas' continued commitment to the highest level of retailing.

ETRO

PAUL SMITH

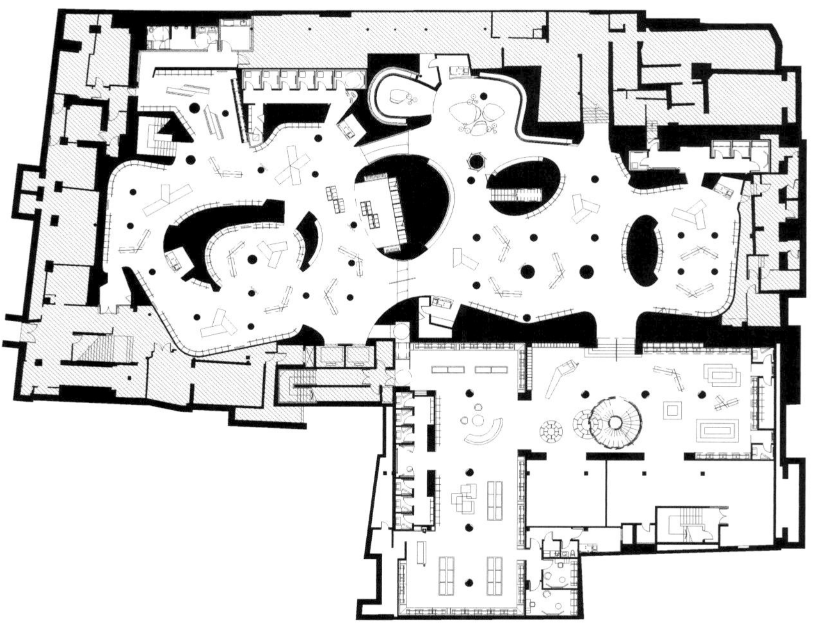

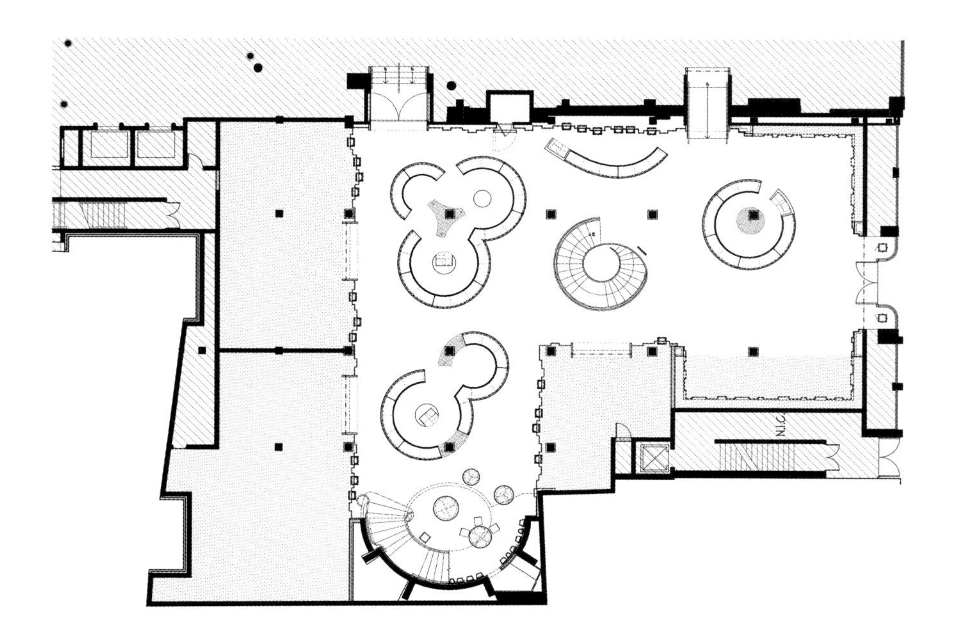

LINKS

Marni Boutique

设 计 师：Simon Mitchell, Torquil McIntosh, Giorgia Cannici
设计公司：Sybarite
公司网址：www.sybarite-uk.com
项目地点：Las Vegas, USA
建筑面积：220m²
摄 影：Donato Sardella

Marni旗舰店的设计灵感源自于一条开裂的鞭子，就像悬浮在半空中一样。这个如蜿蜒绳索一般的不锈钢建筑架构起了这家精品屋，给成品衣系列提供了完美的空间。空间的一端设为收银台，另一端是一堵雕刻墙面，内嵌玻璃展示鞋柜。

曲线形的灰色墙面装饰着一排排的泡泡，凹凸不平，背光阴影，打造了一个富有质感的墙面。内嵌的泡泡里展示着一些小商品，强调唯一性和价值观。其他的墙面都设有背光玻璃展示鞋柜，并随意地放置着一些天然玻璃人体模型，紫光漆光芒四射，这也是受Marni最新配饰的启发。墙后则是试衣间、储藏室和办公室。

独立式的椭圆形展示桌面同样是光芒四射的紫光漆，配以白色凳子及柔软的羊毛垫子，提供舒适的坐椅。天花板上大块的碟片造型与墙上的泡泡遥相呼应，放射出柔和的光线；抛光的水泥地板则提供了一个干净利落的背景幕。同样的，外部设计极具简约。简单的玻璃正面置有一些人体模型取代了传统的展示窗，从外往里看，一览无余，并使中间的展示台脱颖而出，吸引旁边自动扶梯上来来往往的人群。

Sybarite's design for the Marni flagship at the Crystals in Las Vegas was inspired by the image of a cracking whip, seemingly suspended in mid-air as it unfurls. Defining the perimeter, this sinuous "lasso"of stainless steel encircles the boutique, providing hanging space for the RTW collection. At one end it is anchored by the cash and wrap desk, and at the other it morphs into a sculptural wall inset with fibreglass shoe displays.

Painted smooth grey, the curving walls are broken up by an array of bubbles in relief - randomly concave and convex, backlit and shadowed - which build up in textural composition. A selection of accessories are displayed in some of the recessed bubbles, enhancing the perception of value and uniqueness. Other sections of the perimeter contain backlit fibreglass display boxes and scattered throughout are suspended mannequin pieces in a mix of natural fibreglass and pearlescent purple lacquer, a new finish inspired by Marni's latest accessories collection. Beyond the walls is space for fitting rooms, stock room and office.

Freestanding elliptical display tables, also in pearlescent purple, offer additional display surfaces and clusters of white PVC stools and soft grey wool rugs provide comfy seating. In the ceiling, giant Barrisol discs echo the bubble motif of the walls and cast soft diffuse light, while the polished concrete floor provides a clean backdrop. Similarly, the exterior treatment is minimalist. Rather than a conventional window display, a simple glass facade with a few hanging mannequins allows a clear view into the shop, putting the collection centre-stage to draw the attention of traffic from the busy escalators nearby.

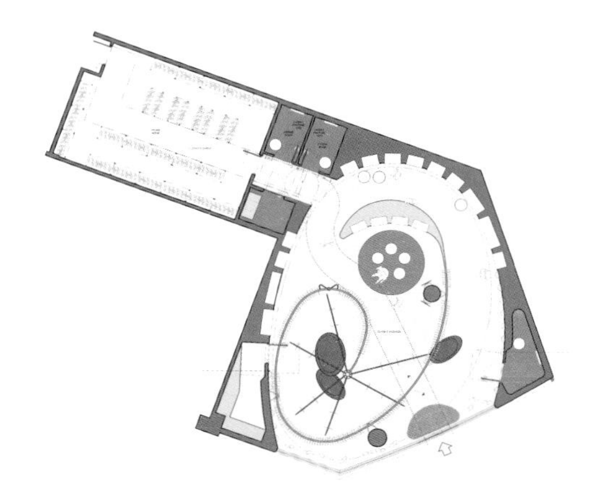

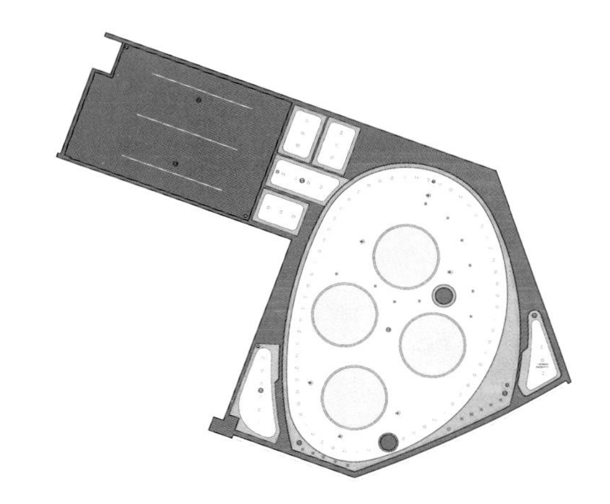

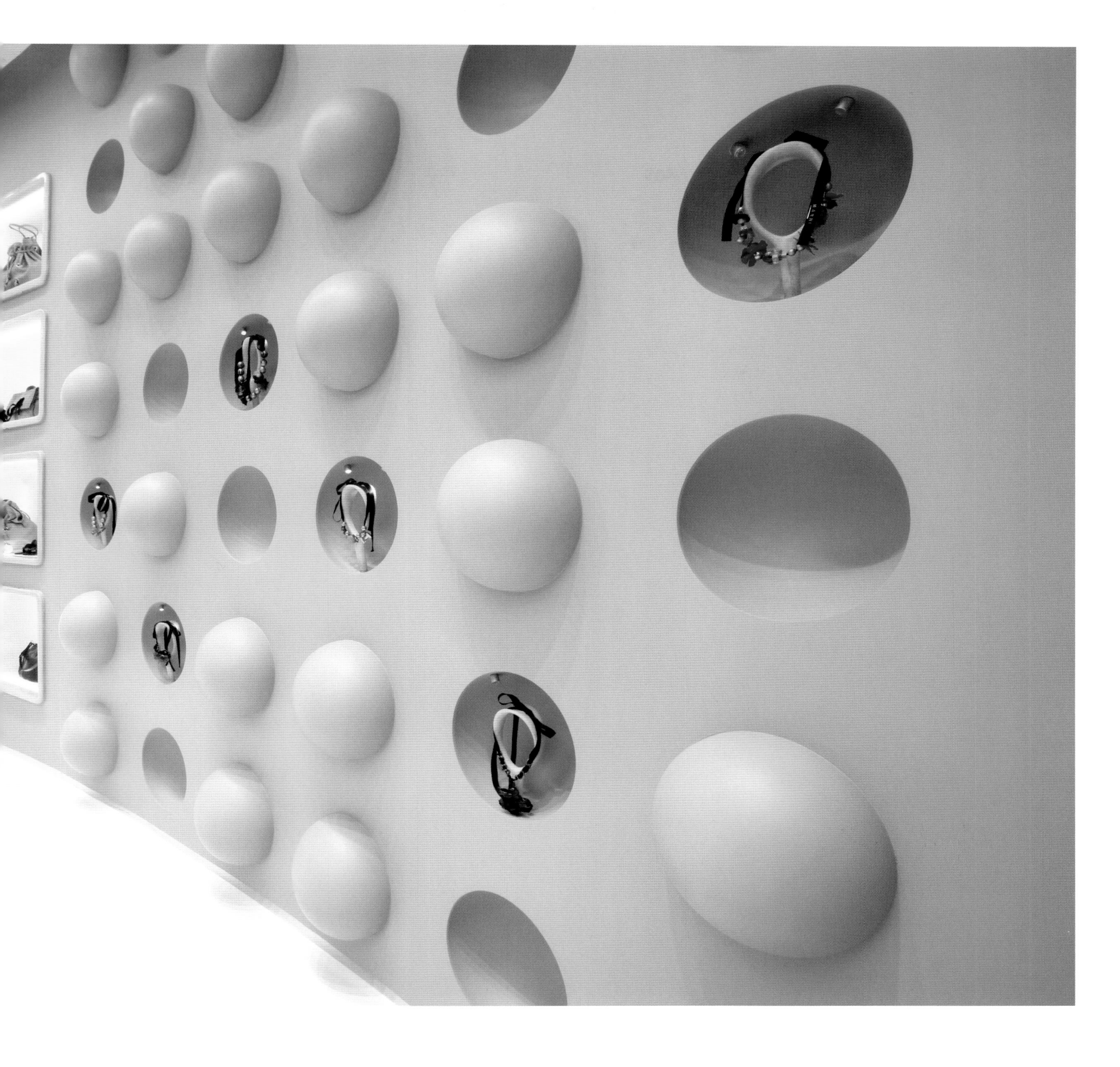

Amicis Women Fashion Concept Store Vienna

设 计 师：Denis Kosutic, Carina Haberl, Alice Cappelli, Judith Wölkl, Matteo Trentini
设计公司：Denis Kosutic
项目地点：Viemma, Austria
建筑面积：335m² Sales room; 267m² Store room
摄　　影：Lea Titz

破损的、几乎空置的大厅似的房间经过十分仔细的处理——拆除的痕迹和新增的设施都以精心设计好的方式呈现出来。这使房间的基本结构显得很粗犷。

敞开的包厢根据“屋中屋”的原则组织和摆放，它们的表层由未加工的实心石膏板制成，使它们能够融入整个空间的结构。内部设计有四个极其雅致华丽的包间，每一个都会因为自身的氛围让人惊叹。它们有着电影摄影棚的效果，与大厅的“赤裸”形成鲜明的对比，展现出了截然不同、非常规的世界——Flower Power, Neo Baroque, Jeams Bond, Boudoir。针对性地使用全新的和古老的家具、织物、灯具，突出了独特的主题世界，使它们拥有一种十分温馨的格调。

考虑到视线被全部装有镜子的更衣室所阻隔，因此并未出现各个空间的直接对话。镜面在更衣室、室内陈列家具和橱窗上的运用，呈现给顾客出乎预料又令人迷惑的视觉效果。同时，这些主体部分通过环境的反射变得“非物质化”。

The gutted and completely emptied, hall-like room was treated very carefully-traces of the demolition and the new installations have been left visible in an elaborate and well-planned way. This leaves the basic structure of the room rough and ostensibly crude and "not completely finished at the first glance".

Released boxes organised and positioned according to the principle "Room in the Room" in their outer skin made from raw, filled plasterboard adapt themselves to the structure of the hall. In the interior, each of the four particularly finely and nobly designed room boxes surprises with an atmosphere closed in itself. They have the effects of film sets in the studio, are in stark contrast to the "nudeness" of the hall and simulate absolutely different and unconventional worlds-Flower Power, Neo Baroque, Jeams Bond, Boudoir. The targeted use of both new and vintage furniture, fabrics and lights underline the individual topic worlds, giving them a particularly homely touch.

The dialogue of these worlds does not emerge directly, given that the visual axes are constantly interrupted by the completely mirrored and equally free-standing cubes of the changing rooms. The consistent and intensive use of mirrored areas, as in the case of the changing rooms, presentation furniture in the room and in the shop windows, allow the beholder to gain unexpected as well as often surprisingly confusing perspectives. At the same time, these bodies "dematerialise" through the reflections of the environment.

AMICIS
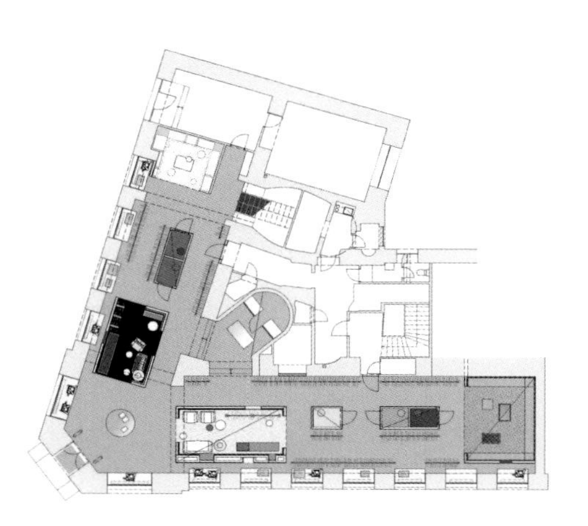

大厅功能性的基本照明和商品的一般照明安装在四个很长的前台位置，灵活多变，极具效果。每个包间的照明使包间呈现出商品和装置灯具的精妙结合。
由于现存的窗口面积大，外表不引人注目的玻璃装置，商店的内部轻易就可以从外部看到。同时，街头生活也很强烈地反映在商店内部，日常生活的都市感丰富了整个房间。

The functional basic lighting of the hall and the general illumination of the goods, mounted on four long on-stage positions, has a provisional effect and is hence also dynamic and changeable. The illumination of the individual boxes presents itself as a subtle mixture of goods and ambience lighting.
Thanks to the large-sized, formally unobtrusive glazings in the existing window openings, the shop interior flows unrestrainedly to the outside. At the same time, street life is being intensively reflected to the inside and enriches the room with its urban scenes of everyday life.

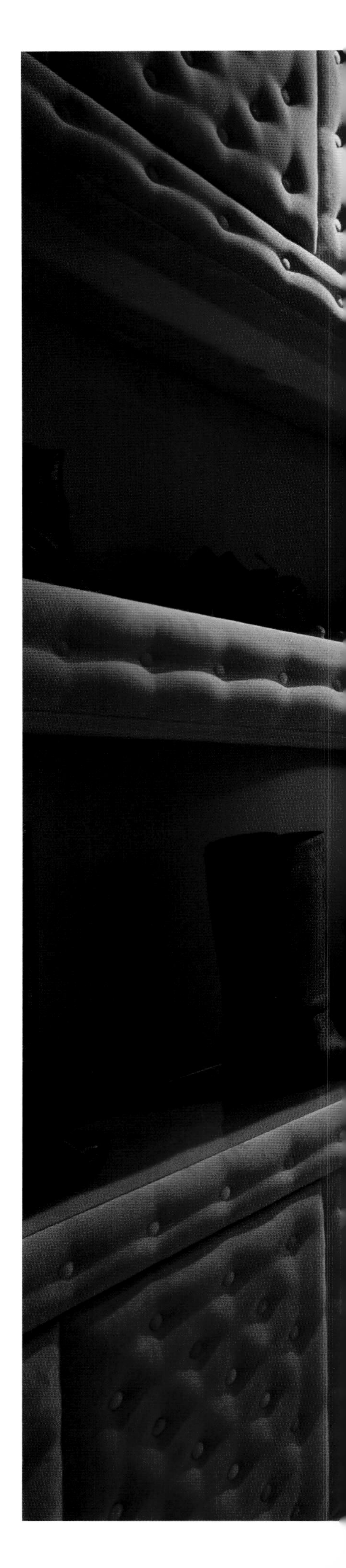

Amicis Bel Etage

设 计 师：Denis Kosutic, Carina Haberl
设计公司：Denis Kosutic
项目地点：Vienna, Austria
建筑面积：$350m^2$
摄　　影：Lea Titz

现实与虚幻的古典元素的大胆融合占据了整个像大厅一样的房间。

二手的古老家具呈现在装饰织物、壁纸、墙面覆盖层上，并赋予了新的定义，形成了展示时尚的高雅舞台。在这儿，由于它们高档的品质，这些服饰陈列在最显著的位置，是舞台上当仁不让的主角。

米黄色和棕色色差的细微差别十分关键，它使房间呈现出微妙的单色效果。因此产生了一种“新式约束”的效果，形式夸张却不乏功能性。每一个刻意的夸张效果在规划阶段就已经被充分考虑在内了，这种夸张效果也是点到即止，不至于超出了自创的范围。

材料的严格挑选，以及它们的图案、装饰和外观，体现了某种立体感、达到一种特殊的、精致的细节渗透。一些技术元素和现代家具创造出了必要的不确定性，从而突出了形状和风格的展示。

现代的宫殿，代表着历史和未来。

A bold mixture of "real" and "unreal" classicistic elements dominates the hall-like room.

The used, antique pieces of furniture blend with newly interpreted quotations of various epochs of style, which assume their shapes as fabrics, wallpapers, stucco, wall claddings and photo collages, forming an elegant stage for fashion. At that point, the dresses - thanks to their high-end quality - stand in the foreground as totally selfevident elements of the mise-en-scène.

The consequential use of countless nuances of beige and brown gives the room a subtle, monochrome efect. A thereby formulated "new restraint" plays, in a pompous manner, with functionalism. Every intended exaggeration has been fully thought out during the planning process, however carried out only to a certain degree so as not to transcend a self-invented boundary.

A precise selection of the materials, whose various patterns, ornaments and surfaces always construe a certain kind of three-dimensionality, enables a special, detailed penetration in detail.

Technoid elements and contemporary furniture, which have been added to the latter, create the necessary uncertainty, thereby emphasising the play of shapes and styles.

A Palace of Today, for Yesterday and Tomorrow.

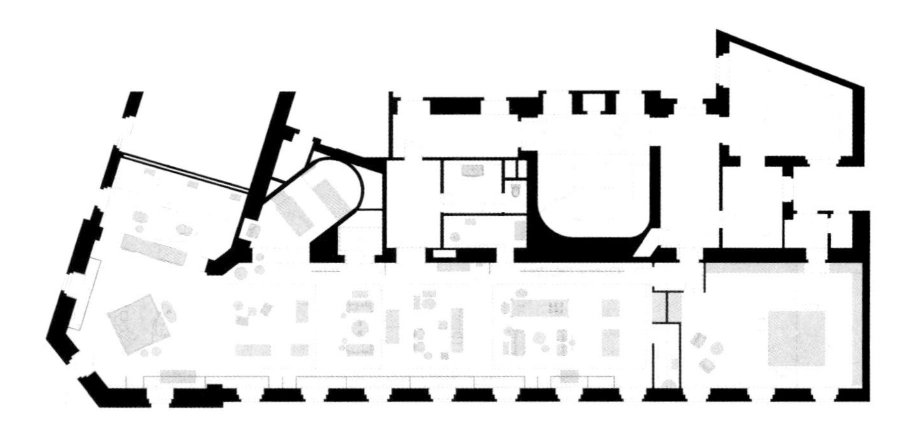

Sigrun Woehr

设 计 师：Peter Ippolito, Gunter Fleitz,
Silke Hoffmann, Judy Haenel, Britta Kleweken
设计公司：Ippolito Fleitz Group GmbH – Identity Architects
公司网站：www.ifgroup.org
项目地点：Karlsruhe, Germany
建筑面积：210m²
摄　　影：Zooey Braun, zooey@zooeybraun.de

Sigrun Woehr是高端鞋类品牌，卡尔鲁厄这家商店标志着Sigrun Woehr的新发展——将其领域开拓到包括新型时装和配件。
新商店将坐落于市中心，其建筑平面狭窄，向建筑内部延伸了将近25米。设计师的任务就是创造出一个可以展示独有商品的空间以及吸引顾客进入商店。
空间的天花板特别设计以创造一个强烈的第一眼的冲击效果。整个天花往后壁逐渐降低，使整个空间更富活力，也使整个空间产生了一种吸引力。环形天花上三幅优雅而时尚的紫色、桃红色和淡粉色搭配出来的图案，创造出了极富吸引力的视觉中心，也将顾客的视线带向了整个空间的尽头。

Sigrun Woehr is the premier address for high-end footwear in the state of Baden-Wuerttemberg. In 2003 we realised the Sigrun Woehr flagship store for our client in Stuttgart. We were now commissioned to develop an interior for the second Sigrun Woehr shop in Karlsruhe. This shop marks a new departure for Sigrun Woehr as she expands her range to include a new line of fashion and accessories.
The new store was to be housed in a shop space in the city centre, which has a narrow floor plan stretching back almost 25 m into the building. Our task was to create a spatial situation in which to present an exclusive range of goods, while at the same time enticing customers across the threshold.
The ceiling of the space was specially designed to make a strong initial impact. The ceiling is gradually lowered over the entire length of the shop towards the rear wall. This gives the room a more dynamic feel and creates a kind of suction pull into the space. Three circular ceiling motifs executed in an elegant and fashionable palette of violet, fuchsia and pale pink tones create attractive focal points and draw the customer's gaze towards the far rear of the space.

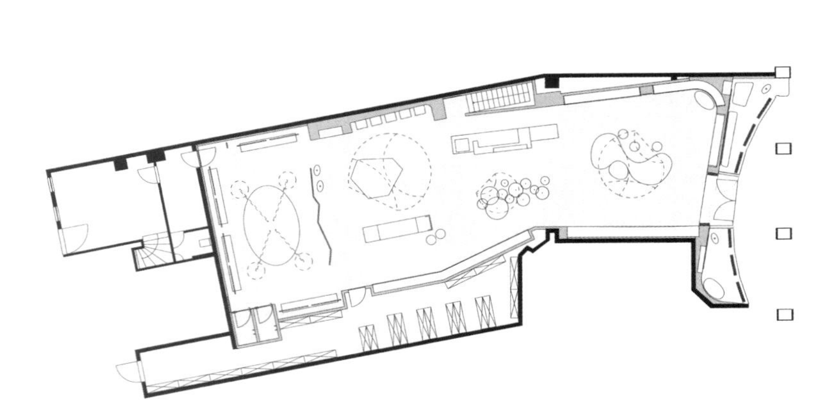

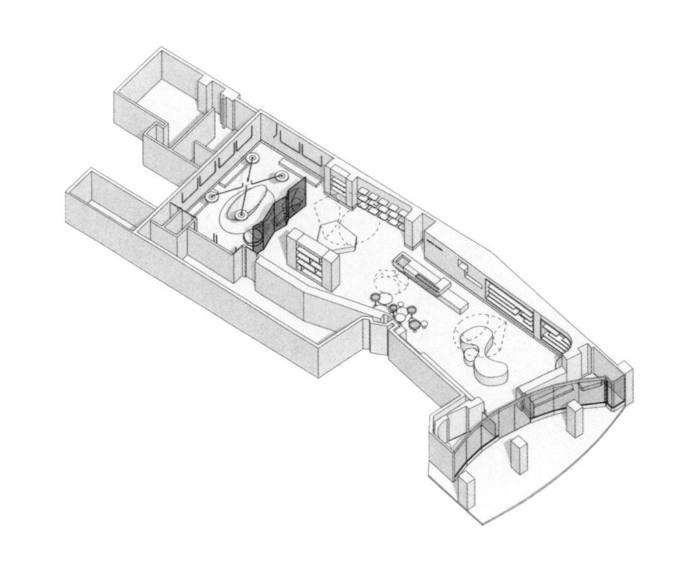

深色的橡木镶花地板营造了一个绵延流动的广阔空间，与天花形成了强烈的反差。每隔一段距离就有一个配件在这样的背景中出现，与天花上的图案相呼应。整个房间分成三个区域，将每个元素置于房间的中央就形成了这个区域的核心部分。不同的几何图形是专门为不同的元素挑选的。入口处一个中央显示器使得商品得到显著提升，它与众不同的形状也能吸引好奇的顾客进入商店。下一个区域由收银台引入，层状的矩形创造出了活动的效果，也为玻璃陈列柜提供了空间。收银台对面是一批圆形的桌子。他们高低、大小各异，表面材料也不同——一些是镶有镜子的，其他是由玻璃或钢铁制成的，而这些也构成了第二处引人入胜的展示区。第三个区域在商店的中央，凭借黑色的活动隔板，两根支撑柱得以隐藏在自立式货架的结构中，货架上有大小不同的展示区。货架的旁边有一个多边形的坐椅以便交流。

A dark-stained, oak parquet floor, which creates a continuous, flowing expanse throughout the space, provides a powerful contrast to the ceiling. Solitary fittings are staged at intervals against this background, each accompanied by a colourful ceiling graphic above. Positioned in the centre of the room, each element forms the core of one of three zones into which the room is divided. Different geometric shapes were deliberately chosen for the individual elements. An amorphous central display unit in the entrance area gives a striking upbeat to the collection. Its unusual form also serves to draw curious customers into the space. The next area is introduced by the cash desk unit. Layered rectangles create a mobile effect and harbour space for glass presentation cabinets. The cash desk unit faces an ensemble of round tables. The varying heights, sizes, and surface materials of the tables - some are mirrored, others made from glass or steel - create a second attractive presentation area. The third area is denoted by two features in the centre of the room. Two supporting columns are concealed in the frame of a free-standing shelving unit containing presentation segments of varying dimensions, thanks to flexible, black separators. The shelving unit is grouped together with a polygon seating element, which also functions as a communication island.

分隔这三个区域的墙壁设计成灰色和米黄色。白色的展示架和壁龛或嵌在墙上或重叠在墙壁内。最为理想的照明方案使所有陈列的商品都呈现出完美的状态。

在商店的最远端有一个专为新型、高级时尚系列而设的独立空间。这个独立空间的入口是两个人形模特，背后是烟雾缭绕的六角形玻璃墙背景。六角形的玻璃给人一种空间一分为二的视觉感受，而玻璃的烟雾也产生了半透明过滤器的效果。

一张长毛绒的粉色地毯上摆着的大椭圆桌占据了这个区域的主要位置。桌子上方的天花板嵌壁式并且装有镜子。四盏安装在突出装置上的纺织灯，从这个开口射出的光线跨越了整个空间。这个区域的隔墙用香槟色的鳄鱼皮材质的墙纸覆盖，为悬吊式挂架创造了一个极具风格的背景。

The walls enclosing these three zones are executed in grey and beige. White presentation shelving and niches are either recessed in the walls or superimposed against a recessed wall. An optimal lighting scheme results in immaculate presentation of all the goods on display.

A separate department was created for the new, high-class fashion line in the far rear of the store. This separate area is introduced by two mannequins positioned in front of a backdrop of a concertinaed, smoky glass wall. The glass concertina visually divides the space and the smokiness of the glass acts as a translucent filter.

The area itself is dominated by a large, oval table that stands on a plush, pink carpet. The ceiling above the table is recessed and mirrored. Four textile lamps mounted on projecting arms emit from this opening and span the entire space. The enclosing walls of this area are covered in a champagne-coloured wallpaper with crocodile leather texture, creating a stylish backdrop to the ascetic, suspended, steel clothing rails.

sigrun woehr

sigrun woehr

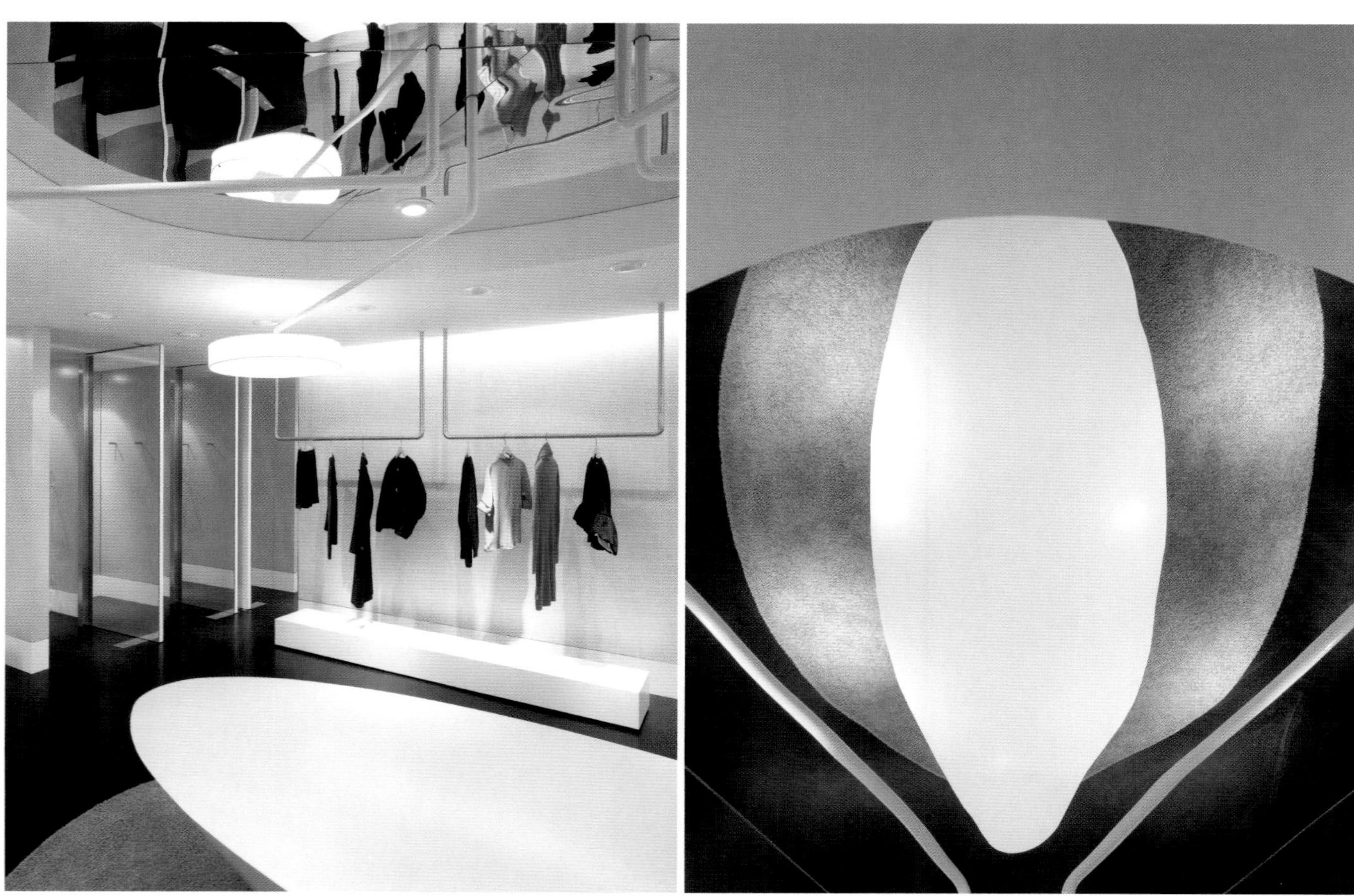

Vakko Nisantasi

设 计 师：Seyhan Ozdemir, Sefer Caglar
设计公司：Autoban
项目地点：Istanbul, Turkey
摄　　影：Ali Bekman

Vakko旗舰店代表了土耳其最奢华的品牌。Autoban面对挑战，选择奢华的材料和量身定制的椅子来装饰全部的五个楼层。受Vakko螺旋形商标的启发，Autoban专门为该项目设计了一个灯的造型，并带有类似弧线形灯罩的顶部，而内部的表面则嵌有简约有序的图形，为整体体验带来无尽的奢华感受。

Vakko flagship store stands to represent the true identity of the countrys finest luxury brand. Autoban rose to the challenge by choosing luxurious materials and customized seating throughout the five floors of the store. Insipred by the swirling Vakko logo, a lamp with a similarly curved lampshade was produced specially for the project, while sleek and simple embedded pattern were introduced along the internal surfaces, adding to the luxurious element of the overall experience.

VAKKO
VAKKO

VAKKO

VAKKO

VAKKO

VAKKO
VAKKO

VAKKO

YAKKO

Jean Claude Jitrois

设 计 师：Christophe Pillet
设计公司：Agence Christophe Pillet
公司网站：www.christophepillet.com
项目地点：Cannes, France
建筑面积：68 m²

本案设计理念由巴黎伦敦玛贝拉演化而来但又包含有了他们的精神，气氛大体一致但又更深层次、更多方面地展示了 Jean Claude Jitrois的性格特征。
厚实奢华的地毯、鼹鼠皮的颜色使客户不知不觉中脱掉了鞋子，纵情地体验着这些精品。服装围栏也是透明的，利用不锈钢的发光树脂玻璃，最大程度地放大了空间的视觉效果。不可置否地说，该精品店既现代又时尚，就像戛纳本身一样，而同时又反映了戛纳永恒的高品质，使之独一无二。黑色的正面映射了璀璨的阳光和克鲁瓦塞特大道上有象征意义的棕榈树。同样的，地下室里的滑门打造了一个VIP空间，若隐若现，既保护了隐私又不会造成视线上的障碍，与楼梯旁边的区域有异曲同工之妙。对于那些专注而有耐心的绅士们，则设有大大的沙发，既现代又富有诱惑力，大而舒适的貂皮坐垫道出了自身无尽的奢华。

The concept has Evolved from the boutiques of Paris, London, and Marbella but keeps their spirit. The atmosphere is the same but goes slightly further in revealing more aspects of Jean Claude Jitrois's personality.
The thick and luxurious carpet, a taupe, moleskin color, invites the client s to take off their shoes and indulge in exploring the collection for themselves. The clothing rails are transparent, using a lightened Plexiglas with stainless steel, witch maximizes the spaces to its greatest extent. The boutique is undeniably contemporary and modern, like Cannesitself, but it also refects the timeless qualities that make Cannes so unique: the sea and the sun. The patent black of the facade refects the shining sun and the iconic palm trees along the Croisette. In the same way, sliding dors in the basement create a VIP space soi t is possible to be seen without being see, conveying the same sense of protection without making a solid barrier, in a similar way as around the staircase. For the attentive and patient men there is a large sofa like in Paris; modern and seductive, with big comfortable mink cushions, which reflect the garments themselves.

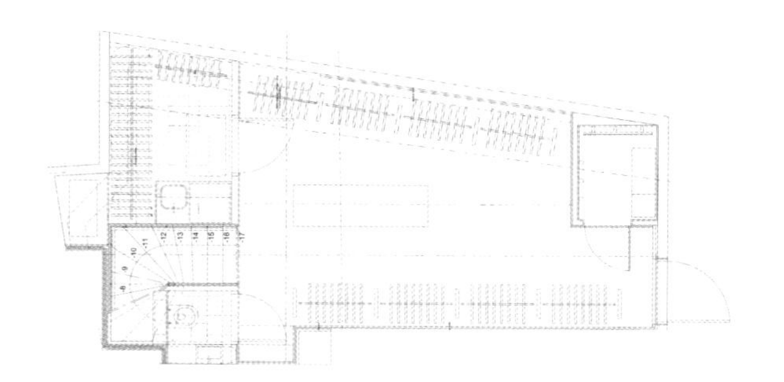

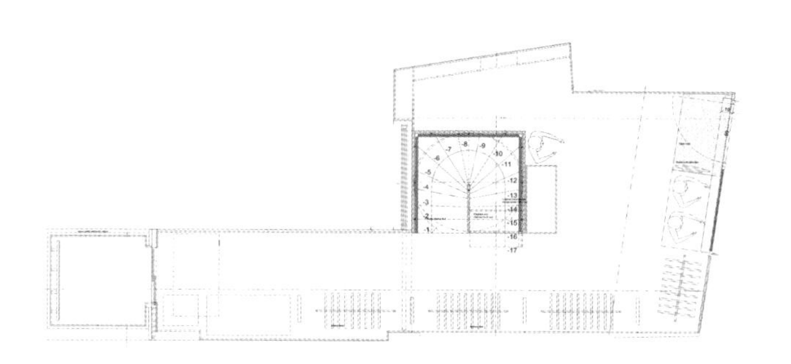

Eger Karl Johan

设 计 师：Vesma Kontere McQuillan, Nichlas Hoel
设计公司：Scenario
项目地点：Osio, Norway
摄　　影：Gatis Rozenfelds

Eger Karl Johan是一家位于奥斯陆中心地段Karl Johan街的高端百货商店。

这是奥斯陆第一个有着明确空间时尚、概念时尚，并强调内部设计的百货商店。

设计的一个最重要的想法是做一个舞台或者“T”台。这个概念贯穿了5栋历史建筑的25个商店，每个都有自己的独特设计。Eger Karl Johan是在奥斯陆历史环境里一个前沿的现代内部设计。

Eger Karl Johan的室内设计以黑色为主调。黑色是典雅和时尚的代名词，因此也最适合用来展示商品。

内部设计另外一个重要的方面是材料的选择，很多都是少而精贵，比如妆点在建筑物之间的意大利的玻璃马赛克。某些材料，比如二楼的罗马马赛克，在挪威史上还是首次使用。

为了给Eger打造一个特殊的身份，在入口区及庭院处都设有设计特殊的家具。家具的设计诠释了Eger的标志。

Eger是一个艺术和建筑交会的地方，在这里，每一个顾客都是表演的一份子。

Opened in May 2009, Eger Karl Johan is a high-end department store in Karl Johan Street in the heart of Oslo.

This is the first department store in Oslo created with a clear spatial and conceptual idea of fashion and an emphasis on interior design.

The idea of a stage or a fashion catwalk was the key idea in the design of the interior; the concept unifies 5 historical buildings with some 25 shops, each with its individual design. Eger Karl Johan is a place with a cutting-edge modern interior within the historical environment of central Oslo.

The colour scheme of the Eger Karl Johan interior is dominated by black. Black was chosen as the colour of elegance and fashion, ideal for showcasing whatever you choose to put in the foreground.

Another important aspect of the interior was the choice of materials, many of which are rare and expensive in Scandinavia, like the Italian glass mosaics that line the connective spaces between buildings. Certain materials, like the Roman mosaics in the 2nd floor, have been used for the first time in Norway.

In order to create a special identity for Eger, special furniture was designed for the main entrance area and the courtyard. The design of the furniture incorporates an interpretation of the Eger logo pattern.

Eger is a place where fashion meets architecture, a place where every customer is a part of the performance.

EGER
EGER

G-STAR RAW
OSLO STORE - EGER KARL JOHAN
LOWER FLOOR

QUINTESSENTIAL
EGER

SAND

FOR MEN
Marc O'Polo
Marc O'Polo
D&G

Pepe Jeans
LONDON
EGER
3 TREDJE ETASJE
Høyer Luxury
2 ANDRE ETASJE
Høyer Woman
Høyer Man
Høyer Jeans
Høyer Trend
Designers by Høyer
Nespresso
1 FØRSTE ETASJE
Tommy Hilfiger
Marc O'Polo
Rizzo
Esthetique
Esthetique
Nina Skarra
Lexington
Apollo
Nespresso
Eger Café
7-Eleven
0 UNDERETASJE
Come As You Are
G-Star Raw
Marc O'Polo

Delicatessen

设 计 师：Dan Affleck, Adam Hostetler, Leo Mulvehill, Guy Zucker
设计团队：Z-A / Guy Zucker
项目地点：Masaryk Square, Tel Aviv
摄　　影：Assaf Pinchuk

每一个时装顾客都明白一个事实，那就是他买的更多是设计而不是材质。而且，每一件衣服过时的速度取决于时尚潮流和季节的改变而不是取决于衣服的质量。因此，投入在设计上的费用要超过投入在材质上的费用。而在建筑上，投入的平衡是恰恰相反的，在材质上的投入要远远超过设计上的。Delicatessen时装店的设计遵循了时装设计的经济逻辑，旨在颠覆这一传统的投入状况。这个商店的设计引入了两个时尚界的主要策略：瞬态材料的运用以及覆盖整个空间的脱节层的概念。

一种很薄的、未加工的瞬态材料——插钉板，整个商店的墙壁都覆盖着插钉板。通过切割、折叠、显示和包裹的处理，这个材料的原有功能发生了转变，用以创造出展示元素、试衣间、桌子和铺面。将插钉板安装在整个高达5米的空间上，再从插钉板的背后打光，这种五金店的粗材料变成了转瞬即逝的、像蕾丝一样包裹了整个空间的裙子。用来插样品钩的网格孔变成了一种服装样式。这些洞既可以用做展示装置，也可以插上样品钩创造出一种立体的结构。

Every consumer of fashion is aware of the fact that he is paying for design rather than material. Furthermore, the pace at which an item of clothing becomes obsolete is dictated by the change of fashion trends and seasons and not by the quality of the item. Therefore the capital invested in design exceeds the funds invested in material quality. In architecture the balance of investments is reversed, the cost of materials far exceeds the investment in design. The design for the Delicatessen clothing store follows the economical logic of fashion design in an attempt to invert this typical condition. Two main strategies were brought in from the world of fashion and introduced into the space: the use of transient materials and the idea of a disjointed layer that would cover the space.

A thin, raw and transient material; the pegboard, was draped over the space. By cutting, folding, revealing and wrapping, the original purpose of the material is transformed to create the display elements, fitting room, desk and store front. By mounting the pegboard on the entire 5m high space, and lighting it from behind, this rough hardware store material turned into an ephemeral, lace-like dress that wraps around the space. The grid of holes, made for hooks, was transformed into a garment pattern. These holes can either act as a display apparatus or they can hold the bare hooks which produce a three dimensional fabric.

catessen

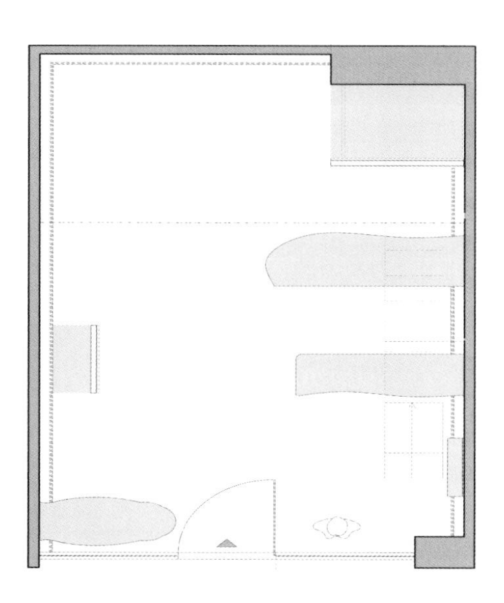

选择插钉板是因为它是基本的活动陈列装置，便于空间的不断变化、发展以及瞬间的改变。空间能够随着展示需要的转变、品牌的演变和单纯的季节的变化而变化。经常光顾的顾客们已经习惯了商品的改变，现在他们就能够享受到更加引人入胜的转变了，而空间设计也变成了一种可以定期消费的商品。

除了垂直的插钉板展示之外，水平位置的展示装置、柜台、夹层和试衣间从插钉板墙上切割出来，从墙上“冒出来”似的，底部都刷成了黄色。这些装置都是重新利用起来的废旧家具（一张老咖啡桌、一张梳妆台和一家小钢琴），它们都漆上了与插钉板相配的颜色。在设计流程的一开始就选择好了材料，这改变了以往传统地将材料应用于概念的设计流程。在这种情况下，材料就显示出了形式和元素功能的各种可能性。

To begin with, the pegboard material was selected because it is the most basic flexible display infrastructure, which allows the constant change, growth and mutation of the space. Spatial transformations can follow a change in display needs, evolution of the brand or simply the change of seasons. The reoccurring customer who is used to the change of goods can encounter an immersive transformation and the spatial design can become a commodity consumed on a regular basis.

In addition the to the vertical pegboard display, horizontal display fixtures, the counter, the mezzanine and the fitting room were cut out of the pegboard dress and "ulled" of the wall revealing the yellow undergarment. The fixtures are found and recycled furniture pieces (an old coffee table, a dresser and a mini piano) that were chopped off flat where they come out of the wall and painted to match the pegboard. Choosing the material at the beginning of the design process reverses the typical design process where materials are applied onto a concept. In this case the materials dictated the forms and functional possibilities of the elements.

North Face

设 计 师：Ken Nisch，Mike Curtis，George Vojnovski
设计公司：JGA
项目地点：Boise, USA
项目面积：107m²
摄　　影：Laszlo Regos Photography

商店位于一栋三层建筑的其中两层。这栋建筑是19世纪90年代建造的，近一个世纪以来都是作为百货商店。建筑的外部得到了必要的翻新，整个外墙涂上了North Face富有乡村风味的红色，与周围同样是红瓦白边的建筑物融为一体。原先的百货商店的遗迹已经转变成了两个独立的零售区，并且重新布局，将巨大的自动扶梯移走，安置一个通风的中庭扶手楼梯。

二楼的窗户从20世纪50年代尘封至今，在其修复中运用了高效的上釉技术，将光线引入内部空间，提供了日光和被动式太阳能供暖。在修复过程中发现了一个建筑的隐藏财富——格栅顶棚。它庞大的面积和技术工艺不仅使内部空间变得温暖，也提高了原本低矮的天花板。其他翻新之处包括原本的砖墙、铸铁的铆钉、金属横梁和围绕着商店的柱子。部分零售区域在二楼，为吸引顾客上楼，设计师采用了突出的视觉元素。顾客一上楼就会看到一幅大型壁画和引人注目的鞋墙。店内还专门设立了一个“社区”，在这儿人们可以分享当地的各种事件和活动，还可以分享自己付出的努力和探索经历。为了一贯的可持续理念，店内还专门设立了一个区域，这个区域里最大的特色就是那些用于回收North Face联合公司产品的专门定制的箱子。

The store is housed on two floors of a three-story building that was constructed in the early 1890s and was occupied by a department store for most of the century. The exterior received a much needed face lift, painting the entire façade a rustic version of The North Face Red to blend in with the surrounding buildings also featuring red brick and off-white trim. The former department store footprint had been converted into two separate retail spaces and needed to be reconfigured with the large escalators removed in favor of an airy atrium staircase.

The restoration of the second-floor windows, closed off since the 1950's, with high-efficiency glazing, brought light streaming into the space, contributing day lighting and passive solar heating to the space. The restoration process unearthed an architectural "hidden treasure", the original wood joist ceilings, whose sizeable presence and artisan-crafted nature contribute warmth to the space and help raise the low ceilings. The openness of the new atrium staircase elements also detracts from the low ceiling height. Other restored elements include the original brick walls, cast iron rivets, metal beams and the columns around the store's perimeter. With the majority of retail space located on the second floor, prominent visual elements were used to attract guests up where they are greeted by a large graphic mural and the dramatic footwear wall. A Community Area was created as a place for guests to share information on local events and activities, along with stories of their personal endeavors and explorations. To continue the sustainability effort, a specific area featuring custom-made bins was designed for the recycling of products by The North Face associates.

ONE WAY
100 N 8th St 200
THE NORTH FACE

THE NORTH FACE
THE NORTH FACE
INSULATION

THE
NORTH
FACE
EQUIPMENT
JOIN OUR TEAM

THE
NORTH
FACE
LIFESTYLE

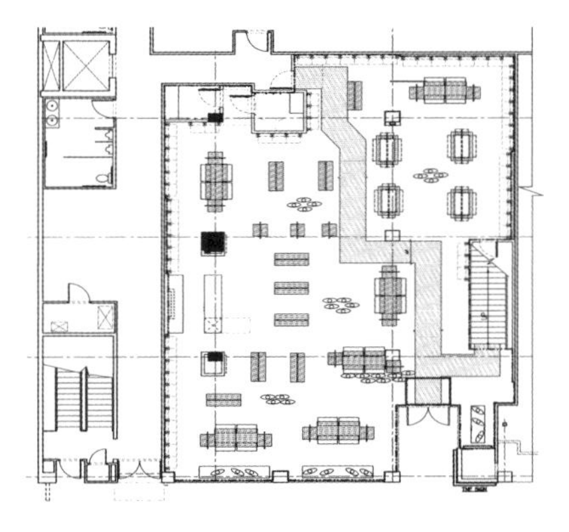

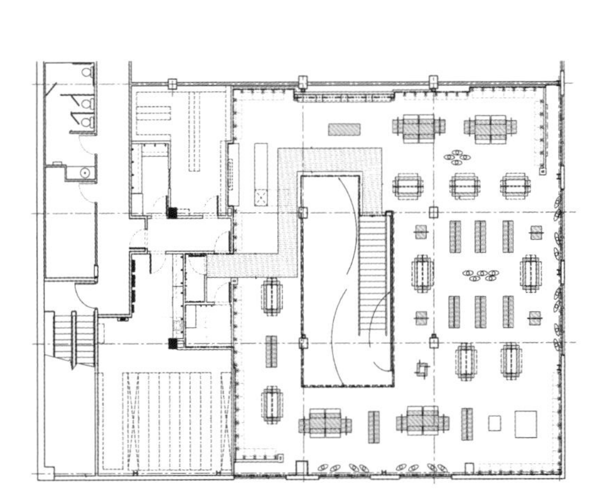

SUMMIT SERIES
GIVE
BACK
THE WARMEST
FULL-BODY SUIT
ON THE PLANET.

GIVE
BACK

STOP EXPLORING
GIVE
BACK

Little Miss Matched

设 计 师：Ken Nisch，Mike McCahill
设计公司：JGA
项目地点：NewYork, USA
项目面积：95m²
摄　　影：Laszlo Regos Photography

这家商店是未来商场、店铺和场内商店的模板。正如这个品牌致力于时尚和乐趣，这家商场反映出了一种“隐藏与揭示”的视角，一种躲猫猫的眼光看待镜子和抛光面上的图案，揭示了每个事物在乍看之下都不明显。

这家商场通过一系列的形状和图形展示这个品牌。这些都极具代表性和功能性，就像在混搭盘朋友们可以相聚，在短袜槽展示出最新的款式以供试穿和混搭。顾客进入商店，有图案的彩色光线投射在入口处的地板上。外部的特点就在于滑动板上，一对垂直的“欢迎光临”（原是商场门前的擦鞋垫）就开始了不对称配对带来的乐趣。橱窗展示时，在插入式平台基座上面的袜子、帽子、手套还有其他商品可以展示在尺寸钉和旋转架上，直立在白色的水磨石地板上。再往里走的区域是成列季节性产品的，例如学生返校服饰、户外服饰，以及夏日海滩上穿的服饰等，这些都体现了这个品牌的显著特点。服饰隔间以墙壁固定装置隔开，创造出商品室。有趣的特点包括一个可以提供建议和邮寄明信片的邮箱，还有相邻的试衣间用他们独特的不符合比例的图形来定义区分。

This Anaheim location is the template for future stores, kiosks and in-store shops. Just like the brand with its focus on fashion and fun, the store reflects a "conceal and reveal" perspective - a peek-a-boo look at patterns reflected in mirrors and finishes that reveal that everything is not evident at first glance.

The store reveals the brand's feel through a series of forms and shapes. These are iconic and functional like the Mixing Bowl where friends can meet, the Sock Troughs where the latest styles are displayed with try-on forms ready for mixing and matching. Shoppers enter the store through colored patterned light projected on the entry floor. The exterior features flip panels that start the fun of mismatching as a pair of vertical "welcome mats". The window display is an inset platform base where socks, hats, gloves or other merchandise can be displayed on dimensional dowel and spinner racks, up-lit from a white terrazzo floor element. The walk-in zone for seasonal products, such as back-to-school or outdoor/summer beach wear, emits the brand's signature style. Clothing bays are divided by wall fixtures to create merchandise alcoves. Fun features include a mailbox to submit suggestions and mail postcards; with adjacent fitting rooms defined by their unique overscaled graphics.

littlemissmatched

eye see possibilities
mixing bowl
try
think
outside
unmatch a batch
kids
3 SOCKS

短袜高柜的白色显示板和弯曲的景观墙延伸到了天花板。那些把商品延伸到天花板的柱子吸引着顾客去探索色彩丰富的混搭选择。有一个圆形的开口嵌入天花板以便混搭盘的柱子可以延伸进去，这增加了比例感。整个建筑的平面布局在销售元素中创造出了弹球效应。前窗和内部镜子表面的分层是由玻璃和薄膜制成的商标。地板上刻着标志性的图案、心形、星星和圆圈，模仿商品的典型样式。墙上贯穿整个空间的大型图案增加了品牌崇尚混搭的气氛。

White display panels on the Sock Tower and curved feature wall have been extended to the full height of the ceiling. The column that extends the merchandise toward the ceiling tempts shoppers to explore the colorful mismatching options. There is a large circular cut-out in the wedge ceiling for the mixing bowl column to extend through, which adds to the sense of scale. The floor plan creates a pinball effect between the merchandising elements.

Layering over the front windows and interior mirror surfaces are logos made of a Lumisty glass translucent decorative film that appear and disappear depending on the viewing angle. Iconic shapes, hearts, stars and circles are cut into the floor to emulate the typical patterns of the merchandise. Oversized wall graphics throughout the space encourage celebration of the brand through mismatching.

mixing
eye see possibilities
eye see possibilities
eye see possibilities
littlemissmatched
any thing goes
bags $20

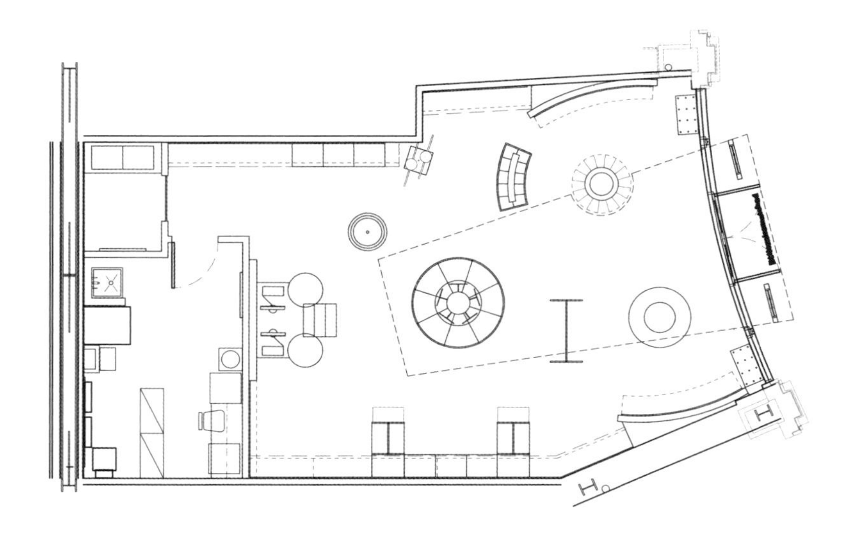

eye see possibilities
littlemissmatched
think
outside
any
thing
goes
kids

sock rocket
3 SOCKS
FEET I YOU

eye see possibilities
littlemissmatched
think
outside
any
thing
goes

Welcome Retail Store

设 计 师：Johan Arrhov, Henrik Frick, Our Legacy
设计公司：Arrhov Frick Arkitektkontor
项目地点：Stockholm, Sweden
建筑面积：160m²
摄　　影：Hannes Söderlund

Welcome是服装品牌Our Legacy和Trèsbien商店合作的产物。

我们尝试打造一种通用的商店理念，这理念可以适用于不同的地理位置和不同的环境中。建筑概念包含了从空间尺寸的分析和展示产品的不同方式中得出的共64种不同种类的家具矩阵。

家具的材料和表现形式可能会因店而异。本案设计包含了Our Legacy的展厅和总部，它是不同愿景的拼接——世界性、私密性、灵活性、物美价廉、独一无二、丰富如诗、俨然城市一角。

不同颜色的材料形成鲜明的个性和强烈的对比，如混凝土、桦木、着色的配线架、塑料、枫木、夹板及钢条都加强了这种特性。

环境会在异物和附近的对比与呼应之间转换。

Welcome is a collaboration between the clothing brand Our Legacy and Trèsbien shop.

Our approach has been to develop a generic store concept that can be applied in different locations and contexts.The architectural concept involves a matrix of furniture, a total of 64 variants that derives from an analysis on dimensions and different ways on how to display the collections.

Expression and materiality of the furniture may vary from store to store. On Sibyllegatan, that also includes showroom and headquarters for Our Legacy, it's a collage of different ambitions - cosmopolitan, intimate, flexible, cheap, exclusive, lush and poetic.A piece of the city.

A material palette of clear character and strong contrasts, such as concrete, birch, stained MDF, plastics, marble, plywood and steel are used to reinforce the ambition.

The idea is that the environment will shift between standing in contrast to and in some cases be fully allied with the store's clothing and accessories.

SOL&
LUNA

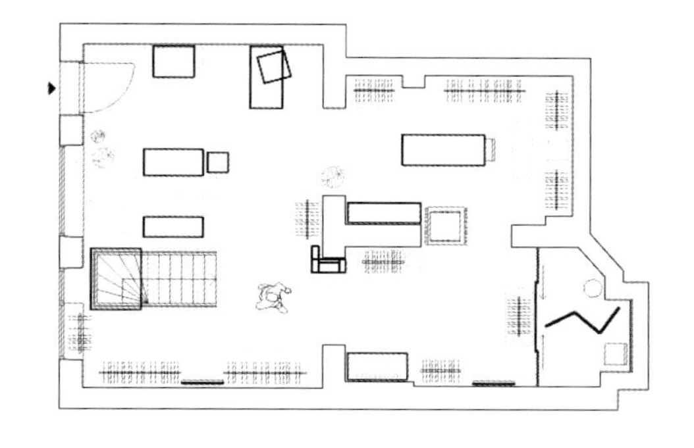

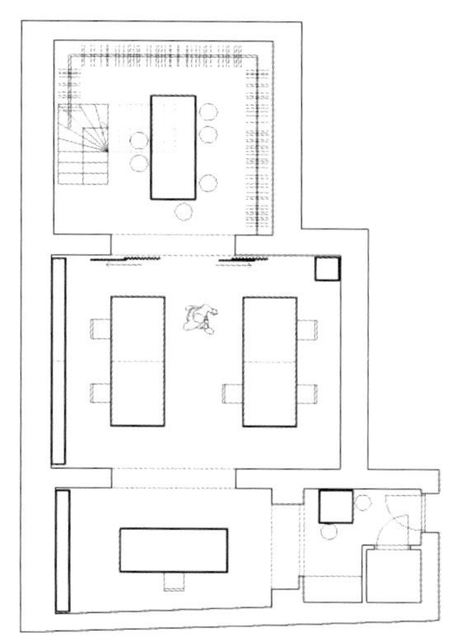

Alberta Ferretti Via Montenapoleone

设 计 师：Simon Mitchell, Torquil McIntosh, Filippo Ferraris
设计公司：Sybarite
项目地点：Milan, Italy
建筑面积：172m²
摄　　影：Stefano Guindani

这家Alberta Ferretti的新旗舰店开设于2010年9月，位于米兰时尚区中心享誉盛名蒙特拿破仑大街。充分尊重这家店所处的现存18世纪新古典主义风格的建筑，不仅是其设计的需要也是出发点。Sybarite的设计方法是还原这栋建筑物最原始的形态，为突出服装的柔美型和轻盈感创造极其简单的背景。定制的磁性显示支架实现了时装陈列的灵活性，而雕刻的独立元素也将Ferretti的品牌印刻在了这个空间中。

原本的外观做了修复，然而内部的门廊依然在使用，这样便将商店前移了，在街上就能够清晰地看到商店内部。现存的两处门廊都简单地上了釉，一个仍然作为出口，另一个变成了玻璃展示橱窗，在它背后安装灵活夹，能够让展示的作品在夜间十分容易就呈现出来，而在白天可以轻松地移走以增加流通面积。面对前门入口处，镶嵌在结构柱上的巨大的LED显示屏上播放着最新的时装秀，而弓形窗口里的定制人形模特就像哨兵一样俯视着顾客进商店。

Opening in September 2010, this new Alberta Ferretti flagship is located on the prestigious via Montenapoleone in the heart of Milan's fashion district. Respecting the existing 18th century neo-classical building in which it's situated was both a necessity and a starting point for the shop's design. Sybarite's approach was to strip the building back to its essential form, creating an extremely simple backdrop to highlight the femininity and lightness of the clothes. Bespoke magnetic display props achieve maximum flexibility in visual merchandising, while sculptural freestanding elements stamp the Ferretti brand upon the space.

The original facade was maintained but the interior vestibule has been utilized, bringing the shop forward and allowing clear views into it from the street. Both of the existing doorways were glazed very simply, one remaining an entrance and the other becoming a display vitrine behind which flexible fixing points set in the floor allow display compositions to be easily created at night and removed during the day to increase circulation space. Facing the front entrance, large LED screens set flush into the structural columns present the latest runway shows while bespoke mannequins set in arching windows look down like sentinels on visitors entering. A previously condemned window facing via Santo Spirito was re-opened, increasing the shop's visibility in that direction.

ALBERTA FERRETTI

依照保护指南，内部结构墙全都保留了下来，它们将空间分隔成独立的房间，靠弓形的门廊相连接。墙壁粉刷成淡淡的灰褐色，地板用白色大理石艺术地板装饰。与现存的房间分隔相一致，艺术地板铺在网格上，其移动接缝自然地形成，并且突出了这栋建筑物的建筑结构。

为了与Alberta Ferretti的理念相一致，灯光调节了情绪使衣服成为焦点。天花遮蔽着背景灯波浪形玻璃，投下奇幻的阴影。人形模特身上的聚光灯给人一种产品自己发光的感觉。标志性的黑色轨道从墙上分离开并且向前悬挂着，同时一张新的“走来走去”的镶嵌着LED屏幕的桌子微微倾斜着，就靠着一端支撑着，似乎违反了地心引力。这样的张力展现出来的能量弥补了其他相对安静、抽象的空间。钢板支撑的磁性半身衣挂和衣架隐蔽在墙里，彻底消失了，这样可以让商品自由地摆放在墙上。

Sybarite利用这栋楼现存的建筑结构和Alberta Ferretti理念的雕刻元素创造出了充满视觉感和大气的空间，衬托了时装系列而没有令它们失色。

In accordance with conservation guidelines, the interior structural walls were all maintained, dividing the space into separate rooms connected by tall arching doorways. Once everything was stripped back to bare bones, the walls were painted pale warm grey and floors finished in white marble seminato. Placed on a grid corresponding to the existing room divisions, the movement joints of the seminato fall naturally and highlight the structural elements of the building.

In keeping with the Alberta Ferretti concept, lighting sets the mood and allows the clothes to become the focus. Barrisol screens back light wave-patterned glass, casting surreal shadows, while spotlighting on mannequins creates the impression that light is emanating from the product itself. Signature black rails are split and cantilevered upwards off the floor while a new "huttlecock" table embedded with LED screen tilts slightly, anchored at its tip and seeming to defy gravity. The energy created by this tension offsets an otherwise calm and minimalist space. Steel panels holding magnetic bust hangers and shelves are concealed within the walls, disappearing completely, while allowing product to be placed freely anywhere on the walls.

Sybarite have utilized the existing architectural structure of the building and the sculptural elements of the Alberta Ferretti concept to create a space which is sensual and atmospheric, complimentary to the collections without overshadowing them.

Lurdes Bergada

设 计 师： Ignasi Llaurad ó , Eric Dufourd, Dorien Peeters
设计公司： Dear Design
项目地点： Barcelona, Spain

这是一个泾渭分明的空间，是Dear Design工作室在巴塞罗那的又一新作。

商店清楚地一分为二，一边用“木质表面”串连起服装店的各个工艺环节，另一边则以服装为焦点。该项目追求保留现有店面空间的共同特性：产业性、差异性、卡通动漫，把当代的东西加入到建筑架构中。

首要的概念是要开拓空间并且可以看得到购物中心的公园。这样会给顾客一种在充满自然之光的地方购物的感觉，即使他们是在一家街头商店里。在此，Dear Design将在购物广场购物的行为解释成“购物大街”。公园也是一个内部的装饰物，背面的外观也以与购物中心内部同样的商业方式吸引着顾客。

A Contrasted Space, this is the new creation of Dear Design studio from Barcelona.

The store is divided into two clearly defined areas which, on the one hand, link every tecnical aspect of a clothes store behind a "wooden skin" and on the other hand, keep clothes as the main focus. The project seeks to preserve the common characteristics of Lurdes Bergada, Syngman Cucala's existing spaces: industrial, different and minimalistic, adding a touch of the contemporary in it's architecture.

The principal idea was to open the space and allow a view onto the shopping mall's park. This would give clients the sensation of shopping in a space full of natural light, as though they were in a street store.Here Dear design apply the main philosophy, interpreting the act of buying in a mall as a "shopping avenue".The park is another decorative piece of the interior. From the park the back façade attract clients in the same commercial way as the interior facade of the shopping centre.

SYNGMAN CUCALA

LURDES BERGADA
SYNGMAN CUCALA

LURDES BERGADA
SYNGMAN CUCALA

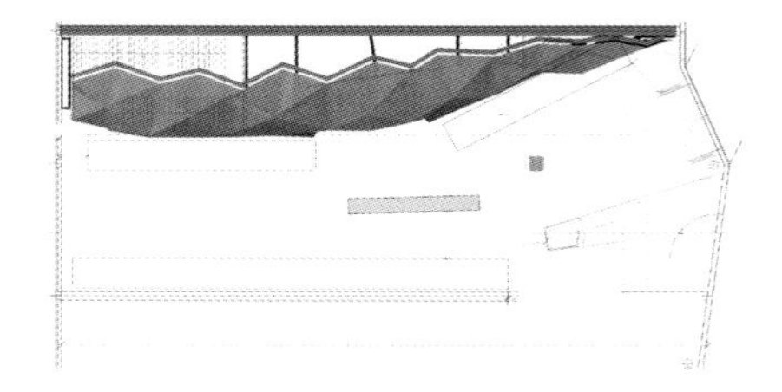

设计团队决定将所有商店的功能都集中在一个大型的独特木质建筑物中，这一建筑物动用了1000片木材和2400只螺丝。木制的外表能够隐藏住所有的大型结构，因此屋顶变得十分敞亮。

从它的外表看，它的“皮肤”向访客展示了它所有极致的建筑工艺的秘密，让人联想到了织布、卷边和针脚等。这反映出了技术在衣服制作中真实的重要性，也增强了这个品牌所传达出的概念：简易、单纯和勤勉。

由不规则的三角形构成的设施显得很牢固，像一个现代的洞穴。自然的山毛榉材质和对面的水泥墙形成了鲜明的对比。

Therefore the design team decided to bring together all the technical functions of the store in a huge unique wood structure made from 1000 pieces of wood and put together with 2400 screws. The wooden skin allows all large structures to be hidden, therefore leaving the roof clear. From it's interior, the "skin" reveals to visitors all of it's extremely technical constructive secrets, bringing to mind fabric, turn-ups, stitching, etc... This reflects the real importance of technicity in clothes making and enhancing the concepts transmitted by the brand: simplicity, purity, and industry.

The installation made of irregular triangles appears rocklike, a contemporary cave. The natural beechwood stands in contrast with the opposite wall, made of concrete.

Levi's Flagship Store

设 计 师：Jeff Kindleysides
设计公司：Checkland Kindleysides
项目地点：London, UK
项目面积：793m²
摄 影：Keith Parry

这家商店通过使购物者体验牛仔布的起源和这个品牌的演变，激发购物者的灵感和吸引购物者，同时提供一种独特的个性化的牛仔服装购买方式。
这家商店设计成一次工匠工作环境的旅途，这个旅途从你踏出街道进入一个“庭院”开始，这个庭院精心设计成展览会一样的空间。白色天花板和再生的砖墙使商店入口变得明亮通风。这个过渡空间被称为“起源”，形成了商店最开始的80平方米。这个空间纯粹是为工艺手段而保留的，它的设计是为了创造一个在充满活力的、迷人的、有创造力的Levi's世界里的经历。“起源”展示从独家产品组合到艺术展览的各种物品，创造出专属于时装零售业的独特的综合性和冲击力。它可以突出新型创新产品同时捕捉想象力，使年轻的创造力和工匠的工作场所相连，吸引顾客进一步探索。
通过两扇巨大的工厂大门，顾客进入商店的主体，这儿展示着最新的时装系列。这儿有着干净和工业的外观感觉，使人联想起车间或工厂。这是有意为之的，也是真实的，特意将它设计成充满能量并且具有功能性。

A demonstration of craftsmanship and factory-inspired architectural design, the store is set to both inspire and engage customers as they are taken on a journey through the origins of denim and the brand's evolution whilst offering a unique and personalised way to buy jeanswear.
Having invented the jean in 1873, Levi's has continued to evolve and re-ignite the category throughout the decades. The brand's rich heritage, inventiveness and understanding sets Levi's apart from any other denim brand. The re-crafted flagship store is designed to demystify what makes one pair of Levi's distinctly different from another - and that of its competitors - whilst easing the buying process for customers.
The store is designed as a journey through an artisan's working environment and starts as youstep off the street into a "courtyard" an area which is crafted to feel like an open, exhibition like space. With a whitewashed ceiling and reclaimed brick walls it provides a light and airy entrance to the store This transition space is called "origin" and forms the opening 80sqm of the store. The space is reserved purely for curation of craft, and is designed to create a vibrant, engaging and creative experience of the world of Levi's, "origin" will showcase everything from exclusive product collaborations to art exhibitions providing a level of intrigue and impact unique to fashion retailing. It serves to highlight new and innovative product whilst capturing the imagination and making a connection between youthful creativity and the workplace of the artisan, compelling visitors to explore further.
Through two sets of huge factory doors, visitors enter the main body of the store where latest collections will be displayed. There is a clean and industrial look and feel, reflective of a workshop or factory. It's deliberately purposeful and real, meaningfully designed to be robust and with function.

Levi's
Levi's

Levi's
Levi's

更多的商品展示在低层的展示架上。在这儿，金属调色板和二叠分的橡木块创造出了一种工厂装货间的感觉。这些调色板和木块建成一系列的矩阵，创造出层列式的展示架，配以商店四周摆设的人形模特，这些展示架讲述着核心产品的故事。在旋转货架的周边和衣架上展示着、悬挂着或折叠着产品，库存活动影像和帆布画框很好地呈现了产品，高水平地突出了存货空间。

一个现代化的楼梯通向地下室，楼梯最大的特点就是每一级台阶上都有一个背光玻璃竖板上面刻着Levi's“XX”的镭射标志。

在令人印象深刻的5米长的收款台的末端，一直到商店的尽头，是501牛仔裤货仓。这个货仓通过上釉的地板到天花从商店中分隔开来，而且设计成装有玻璃的背墙，给人一种似乎这个501牛仔裤拱顶是无边无际的感觉。

Levi's旨在通过这家商店提供给顾客终极的品牌体验。商店在视觉上充满吸引力，提供专门的知识，还有产品和故事。

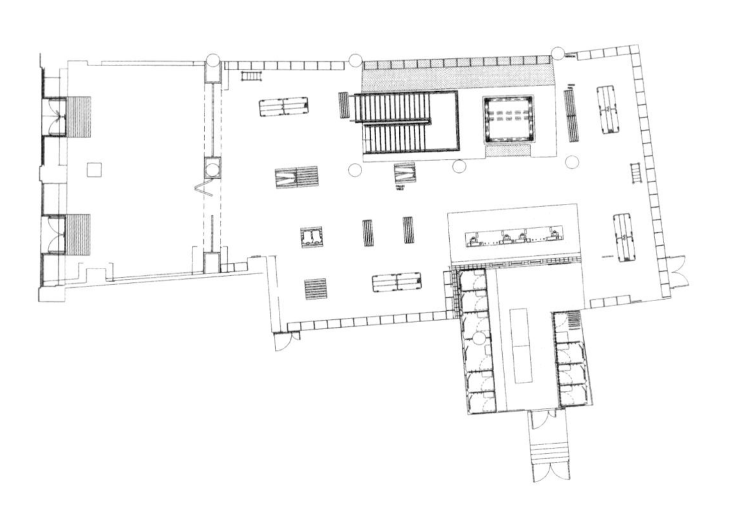

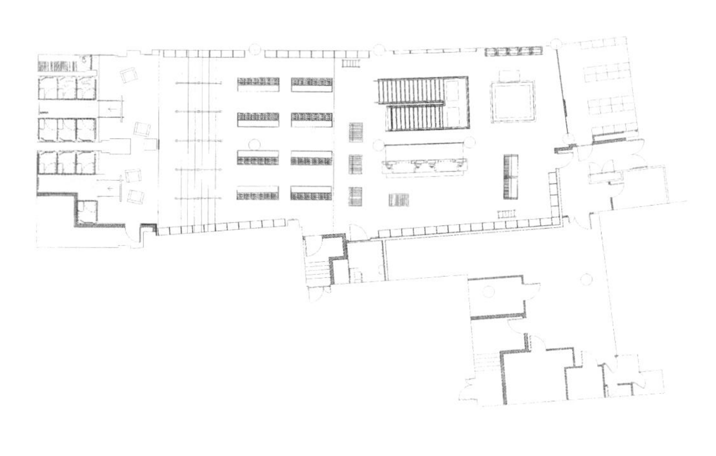

More product is showcased on the lower level displays; here metal palettes, with stacked oak blocks give the feel of a factory loading bay. These palettes and blocks can be built into a range of matrices to create tiered displays. Supplemented with mannequins, which are used in abundance around the store, these displays tell key product stories along with oak and bonded glass display cases for accessories. Around the perimeter rolling racks and shelving display hanging and folded product, whilst seasonal campaign imagery and duck canvas frames presenting key products, punctuate the stockholding space at high level. A contemporary staircase leads down to the basement and features backlit glass risers with Levi's "XX" laser cut into each tread.

At the end of an imposing five metre long cash desk that follows to the back of the store is the 501® Jeans warehouse. The warehouse is separated from the store by floor to ceiling glazing and has been designed with a mirrored back wall to give the impression of a seemingly never ending 501® Jeans vault.

Levi's has created a place where craftsmanship and authenticity deliver the most genuine experience of the brand in Europe. The store's prime position on Regent Street, one of the pivotal and premiere shopping destinations in the UK, will provide Londoners and the capital's visitors with the ultimate opportunity to an engage with Levi's in a unique way.

Timberland Flagship Store

设 计 师：Jeff Kindley
设计公司：Checkland Kindleysides
项目地点：London, U.K.
建筑面积：$238m^2$
摄　　影：Keith Parry

Timberland给了CK一个挑战：要将Timberland的树形商标运用于实际中，用行动体现他们的环境价值观。

从Timberland的商标和形成中心建筑风格的树形顶板支护中得到启发，我们利用一个回收的木材树枝制作格架，将整个商店包裹在这个品牌标志性的商标中。整个外观表达了一个强烈的品牌宣言，钢丝切割的钢铁标志仅仅是为了说明这就是Timberland。

商店的结构使商店内部吸引着人的视觉，不规则的展示窗使每一双鞋子都能够展示在简单的结构里。在商店前部的橱窗里，展示架摆放在精心制作的家具和道具里。

入口大门以废弃的木板构建，欢迎着顾客进入商店。商店前部的玻璃罩和石板的展示柜里展示着最新的Timberland的鞋子和服装系列。商店上面是再生的横梁，既刺激了消费又使4米高的商店降低了焦距。238平方米的商店是以性别分类服饰的，以图形标示，女装在商店的左边男装在右边。商店的中心是“交流图腾”，用于讲述Timberland是如何帮助和支持社区及环境规划的。

As part of the brief for their new store at Westfield London, Timberland challenged CK to bring the brand's iconic tree logo to life and show their environmental values in action.

Taking cues from the Timberland logo and the dynamic tree-like roof supports which form the architecture of the centre, we created a lattice of reclaimed timber branches that wrap the store in the brand's symbolic logo. The façade creates such a strong brand statement that the fret cut steel signage merely acts as endorsement that this is Timberland.

The structure creates interesting views into the store and the expanse of unusual shaped display windows allow almost every item of footwear to be showcased in a simple framework. While in the windows at the front of the store, displays are set against crafted, repurposed furniture and props.

The 3.5m doors, constructed out of salvaged planks, open wide to flank the entrance and welcome consumers into the store.

At the front of the store glass and slate topped display tables showcase the latest footwear and clothing ranges from Timberland.Above, a reclaimed beam allows for intriguing merchandising and lowers the focus in this 4m high store.The $238m^2$ store is navigated by gender, signposted with graphic imagery, with womenswear located on the left of the store and menswear to the right. To the centre of the store is the "community totem" dedicated to telling how Timberland helps and supports community and environmental projects.

Timberland
Timberland

EARTHKEEPER
COLLECTION

AUTHENTIC
EXPEDITION.
WATERPROOF

DESIGN YOUR OWN BOOT
at timberlandonline.co.uk

GAMES
LONDON·1908
DIPLOMA
OF MERIT
WHAT USED TO BE ON THIS SITE?
AWARDED TO
FOR

STORE PAINTER
STEPH KENT
LEICESTER
HAD THESE
SHOES FOR
10 YEARS

WE'RE REMOVING
SOLVENT BASED ADHESIVES
FROM OUR PRODUCTS
SO FAR WE HAVE
PLANTED 700,000 TREES
AND WILL KEEP PLANTING
EARTH
KEEPERS

Earth Market

设 计 师：Yusaku Kaneshiro , Mitsuru Komatsuzaki
设计公司：Yusaku Kaneshiro+Zokei–syudan Co.,Ltd
项目地点：Tomisato, Japan
建筑面积：100m²
摄　　影：Masahiro Ishibashi

室内设计的构想是使用木材和钢铁创造基本氛围以提高温暖和力度的感觉。枝形吊灯的添加产生了更加西式的影响。天花板上的钢丝网使设计变得精致却不是功能性——墙的上部是存货空间可以用来展示“悬挂的”商品。

This original interior was created using wood and steel as a basic atmosphere to enhance a feeling of warmth and strength. The addition of a chandelier produced a more western influence. The steel mesh on the ceiling enables a sophisticated yet functional design - the upper part of the wall has stock storage space and "hung" products can be displayed.

EARTH MARKET

FULL COUNT
Johnbull co.,ltd.
Pherrow's Sportswear
EVISU
Sugar Cane
STUDIO D'ARTISAN & SA.
AMERICAN VILLAGE OSAKA
Skull Jeans
The Flat Head
WAREHOUSE COMPANY
Schott NYC
Ted Company
BARNS OUT FITTERS
nudie jeans
Dry Bones
EARTH MARKET
EARTH MARKET

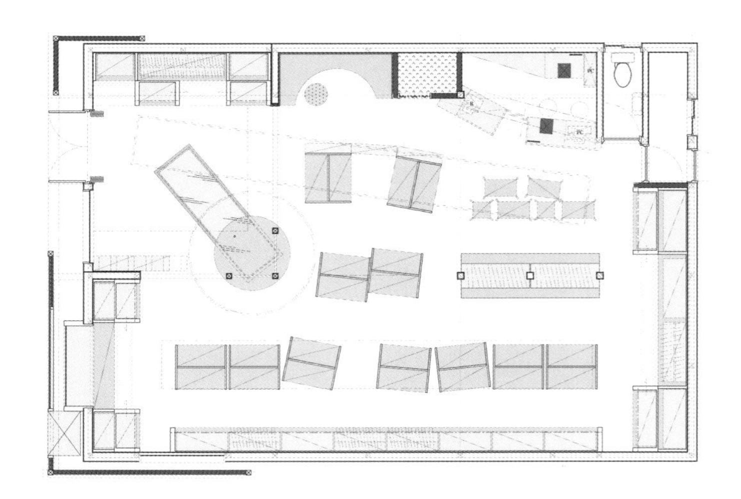

TURNING POINT
STUDIO
D'ARTISA
SINCE 1979
Schott

EARTH MARKET

EARTH MARKET
SINCE 1979
MILWAUKEE
THE FLAT HEAD

EARTH MARKET

EARTH MARKET

FILA

设 计 师：Michael Chan（陈树良）
设计公司：JBM Design
项目地点：中国香港

FILA 是世界知名的意大利运动品牌。品牌一向注重风格的塑造，并以意大利独具匠心的技艺而著称。在服饰创作上，产品拥有深厚的专业体育传承，同时拥有深远的运动生活诉求。具体设计风格是：充满生气，活跃，简洁而着重细节。

我们设计的中心思想，就是引用意大利“金色四边形”的地块及路线图，通过极简的形体、线条展示在店铺中，构建起了一场跨越时空的对话，充分地展现了品牌一向追求的创新精神，更表现出FILA 拥有深厚意大利文化背景的意义。同时，体现品牌文化与特点，“尊重传统，挑战常规”。

FILA is Italian sports brand well-known in the world. The brand always pays attention to shaping style and is famous for Italian great originality technology. With a strong professional sports heritage of clothing and products making, FILA pursues sport life of long history. Specific design style: dynamic, active, simple and detail oriented.

Move to the main idea of our design this time, which is to quote the area and route of Italian Golden Quadrangle by simple shape and lines displayed in the store. Form conservation across space and time, fully reflected the innovation spirit pursued by the brand, besides it also reflected the profound Italian culture background of FILA. Meanwhile, showed the brand culture and characteristic respect tradition, challenge convention.

FILA
EXIT 出口

FILA

FILA
FILA

FILA

此外，设计另一目的是为了创造一个充满活力、时尚的室内空间。针对中国不断进步中的、接受西方事物的一群新消费力量，要突出FILA的个性及与其他竞争对手有着鲜明的分别，使客人每一次光临都有独有的视觉效果及体验。所有坐椅、展示台、中岛架，甚至墙身层板、配件都是活动式，便于多样化的组合。
设计以色彩为创新精神，做出新的注解，纵横交错的蓝色带子作为视觉识别系列的一大元素，结合品牌色系，不断地出现于空间墙面，反复却又富于变化，加强空间整体感的同时也突出了品牌印象。身在其中，如同穿梭于动感的都市，带来了活力的体验。

Besides, the other aim of the design is to create a dynamic and fashionable interior space. Aiming at the new progressive consumptive force that accept western things, and highlight the individuality of FILA and his extinguished difference from the rivals, the shoppers can enjoy unique visual effects and feelings for each presence. All the chairs, platforms, alcatraz frame and even wall layers and accessories are mobile and convenient for diverse combination.
The design uses the color as innovation spirit to note in a new way. As one important factor of visual identity serials, the crossed blue tapes appear on the space wall in combination of the brand color scheme, repeated but full of changes which enhance the sense of wholeness of the space as well as striking out the brand impression. In the store, it seems that we are going in the dynamic city, bring the vigor experience.

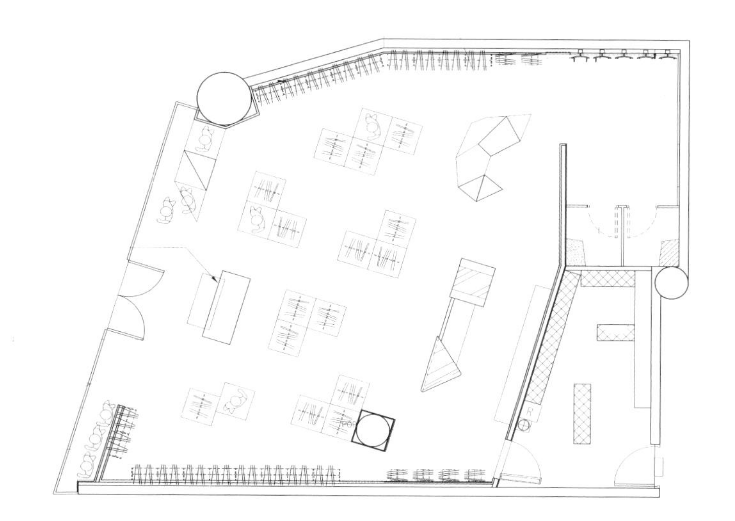

主色调的白色为店铺提供了一个柔和的气氛之余，亦增添一份优美、明亮跳跃的感觉。空间中用上不同类别的物料，以各式各样的材质产生一系列的变化，店里的道具与墙身以FILA 品牌的基本色调——白色、红和蓝色配搭，使它们拥有强烈的视觉冲击力。

The white in the main colors provide a soft and warmly welcome atmosphere as well as adding a feeling of grace, brightness and vigor. The usage of different types of materials results in serials of changes by various textures. The props and walls in the store are matched by the basic color of while, red and blue of FILA which generates strong visual impact.

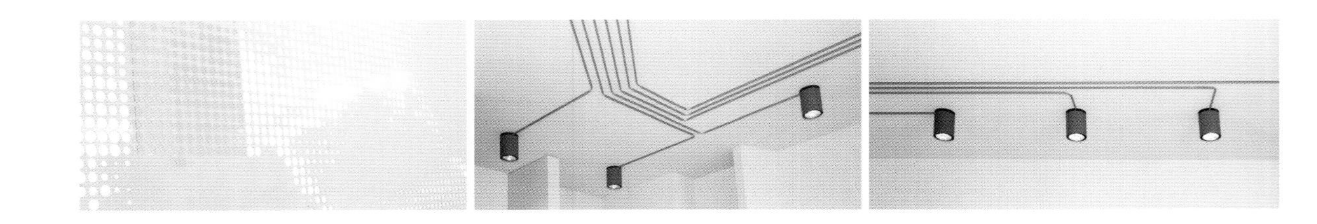

Maygreen

设 计 师：Martin Jacobs, Karim E1–Ishmawi,
Christopher W.Middleton
设计公司：Kinzo
公司网站：www.kinzo–berlin.de
项目地点：Hamburger, Germany

全部由绿光组成的蚱蜢图案装饰了商店的墙壁。精心挑选的时装精品陈列在低矮的多边形展台上，展台就像冰山一样漂浮在深色的橡木地板上。为了创造出蚱蜢的效果，Kinzo特别运用软件将图片的灰度等级转变成许多被激光切成金属片的孔洞。精细的蚱蜢脚和触须都是由开放式布线和绿色的灯光装置反射出来的。“我们将所有的电缆都缠上绿色的尼龙网布，然后把它们固定在天花板上以模仿电路板或是电脑芯片的图案。”Karim El-Ishmawi说到，“我们将电缆作为一种很吸引眼球的设计元素而不是把它们隐藏起来，而这种效果也因为商店的配色方案而得到了加强。”每一条电缆末端都有一盏吊灯或安装好的灯饰。其中一些灯是专为Maygreen特别设计和定制的。

The image of a grasshopper, solely composed of green light, adorns the back wall of the store. Selected fashion items are presented on top of low, polygonal platforms that seem to float on the dark oak floor like icebergs. To create the grasshopper Kinzo specially commissioned software to convert the greyscale of a picture into a pattern of holes that were laser-cut into a metal sheet. The grasshopper's delicate legs and antennae are echoed by the open cabling and green light installation. "We covered all cables with green nylon mesh and fixed them on the ceiling to resemble the graphic pattern of a circuit board or computer chip", says Karim El-Ishmawi. "Instead of hiding the cables we used them as an eye-catching design element. This effect is enhanced by the shop's otherwise muted colour scheme." Every single cable ends at a pendant or mounted light, some of which were individually designed and manufactured for MAYGREEN.

Diesl Denim Gallery Aoyama

设 计 师：Makoto Tanijiri
设计公司：Suppose
项目地点：Aoyama, Japan

牛仔布以往时常被认为是工作服，现已经以各种不同的形式向人们展现了它确实是时尚元素。相同的，一组通常会被忽视的水管设备，却以“自然工厂”的名义展示出了截然不同的表现形式。复杂的水管装置沿着墙体向四面八方蔓延开，覆盖了整个空间，它就像是一棵老树，生长了很久。整个空间覆盖着人造的水管，创造出仿佛是自然乔木的氛围。
水管和时尚饰品营造出了新的景观，展示出这些主要功能明显的东西。事实上空间变得更加多样化，也具有更高的价值。

Denim as recognized work clothes formerly had, at times, shown different expressions as fashion items to the people.?Equally, a group of plumbing, usually unnoticed, shows completely different expressions under the name of "Nature Factory". The complex plumbing, trailing by the wall in all directions will cover all over the space. It is like a tree grown over a long time. An atmosphere like a natural arbor is created in the space covered with artificial plumbing. New attractive scenery is presented with plumbing and fashion-items to show such primarily functional things actually are more diverse and also have higher value.

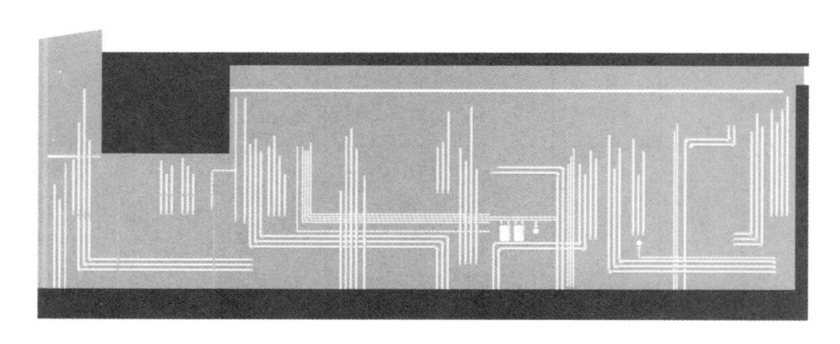

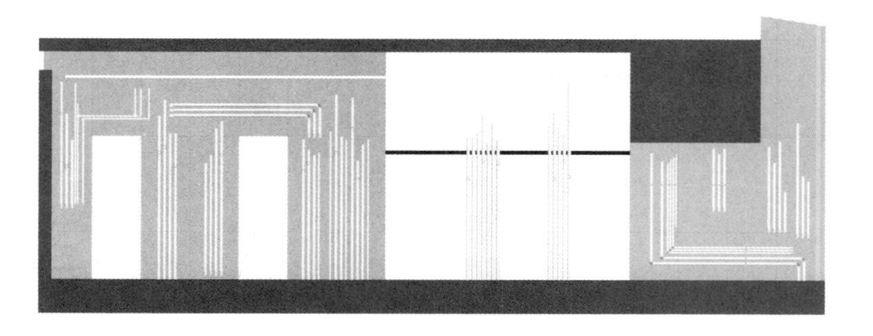

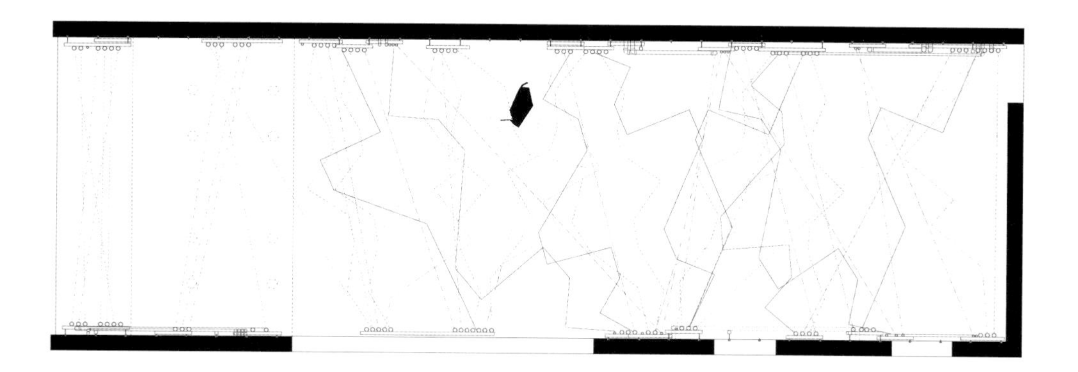

Uniqlo Megastore

设 计 师：Gwenael Nicolas
设计公司：Curiosity
项目地点：Tokyo, Japan
摄　　影：Nacasa & Partners

坐落在东京新宿的优衣库是由设计公司 Curiosity 所操刀设计。这座新的购物广场最大的特色就是入口处的三座显示塔，创造出一个缩微的城市商区。新的设计强烈受到东京街头景观影响，由三个显示塔插入发光的地板创造出一个大型的入口。

入口处以三座巨大的展示塔重现了一个迷你的新宿缩影。随着发光的地板以及明亮的展示塔照亮了附近的街道，这个大型商店变成了一个指标性的地标：吸引人、可靠以及安全的。新的优衣库大型商店在它的商业用途里注入了社会层面的元素。

整个店面的布局让角落与建筑本体的界线变得模糊，而在新宿的繁忙街头塑造了一个独特的氛围。

Curiosity completed the design of the Uniqlo megastore that opened in the station of Shinjuku in Tokyo. The new megastore is symbolized by the 3 display towers in the entrance creating a micro cityscape. The new design is strongly implemented in the Tokyo urban landscape.

The large entrance created with 3 displays towers inserted into a lighted floor, mark the entrance. The lighting floor wrap the towers of a glow of light that lightens the surrounding streets, the shop become an active element of the street topology: attractive, reliable, and secure. The new Uniqlo megastore adds a civic dimension to its commercial purpose.

The layout and angulations of the towers blur the boundaries between the street and the retail space, creating a unique environment in the busy street of Shinjuku.

ユニ
クロ
UNI
QLO
UT
MEGA
CULTURE

UNI
QLO
BRAND SHOP
質
men's FIM

UNI
QLO

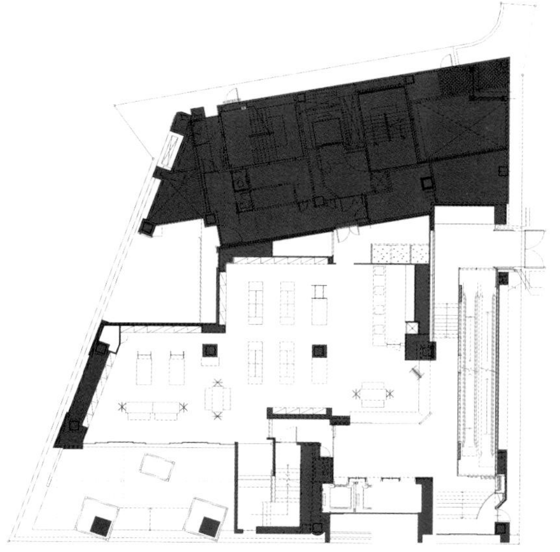

当你在三座塔楼间自由穿梭时，似乎看到了这座城市的缩影。垂直的入口显示塔发射镜面墙上，呈现出类似画廊的惊人效果。这三个交错镜射的展示塔像迷宫似的，似乎整个显示塔一直延伸到了商店内部，营造出塔中塔的感觉。

外观设计延伸到了建筑物的外部，上面全是由几何线条和优衣库的标志组成。

光线变成了抽象的线条，在新宿的夜景中凸显了东京未来的景象。似乎整个建筑的外观形态和建筑本身都消失了，只有几何的线条框架留下了。在夜色中，优衣库商店似乎变成了一个几何图形，仅仅是添加在夜色中的东京这一霓虹迷宫上的一个几何图形。

The city seems to have become human scale as you move freely around the 3 towers. The vertical displays of the entrance are reflected on the mirrored wall creating an amazing gallery of displays, a maze of reflections of reflections, the tower seems to continue within the interior of the shop.

The façade design extends to the building exterior with a composition of graphical lines and Uniqlo logos.

The light become abstract lines in the Shinjuku nightscape emphasis a vision of Tokyo as a future, the materiality of the fa?ade and architecture disappear only the graphical frame remain. At night the Uniqlo shop become just graphics adding to the complex maze of neon and information of the virtual city that Tokyo becomes at night.

ユニ
クロ

UNI
QLO
ユニ
クロ
MEGA
CULTURE
BRA TOP

UT
UT
13
14
15
16

马可玛瑞

设 计 师：迫庆一郎
参与设计：原信敏、岩佐透
设计公司：SAKO建筑设计工社
项目地点：中国北京
建筑面积：89平方米
摄　　影：広松美佐江

马可玛瑞既是儿童时尚服装店，又是作为旗舰店品牌的设计。根据品牌的风格，客户要求设计出欧式风格。基于这样的请求，我们选用了已被欧洲古典建筑定性化的弓形结构作为主题，再把现代风格的装饰融入其中。

这样一种全新的设计理念就出自孩子使用的黏土工艺和像棉花糖一样的圆润造型。覆盖在弓形结构表面的石膏有着细致纹理，其质感给人一种糕点似的柔和的感觉。

我们的设计理念就是把拱形所拥有的力道和童装品牌的可爱，这两个概念融合在一起。拱形在店铺四周连续性地展开，继承了古典建筑中对称的结构。这些连贯拱形环绕的中心地带，是父母和孩子们放松休息的地方。来这里的人们看到的似乎是一个立在庭院里的亭子。无论是休息区还是中岛家具，都重复着拱形和棉花糖似的装饰，而我们正是用这种理念把整个店内的设计风格统一起来。

Marco Mari is not only a children's fashionable clothing shop, but a design for flagship store brand. In line with the brand style, customer demand Europe type design style. Based on the demand, we choose the theme of qualitative arch structure of European classical architecture, blending in the modern style decoration.

The brand new design concept is inspired by the clay craft used by children and rounded design like cotton candy.The plasters covered over the arch structure are of close grain, striking the lookers with a soft sense of texture like a kind of pastry.

Our design concept is to merge the strength of arch structure and loveliness of children's garment brand together. The arch spreads all around the store continuously, inheriting the symmetrical structure in classical architecture. Surrounding by the continuous arch, the central area is the place for parents and children to rest. The comers seem to see a pavilion standing in courtyard. In the rest area as well as alcatraz furniture, the arch and cotton-candy like decorations are repeated, by the concept of which, we unify the design style of the whole store.

marco&mari

marco&mari

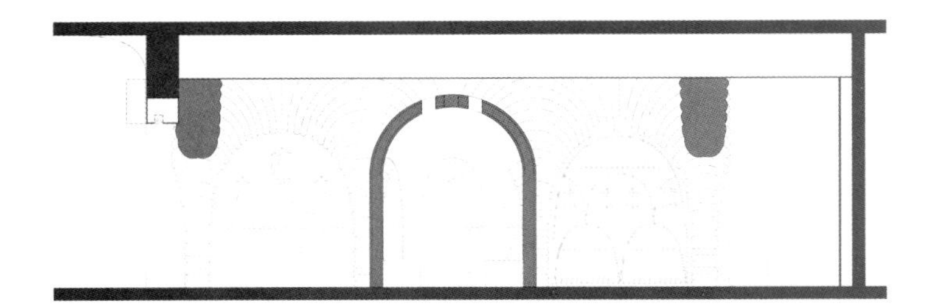

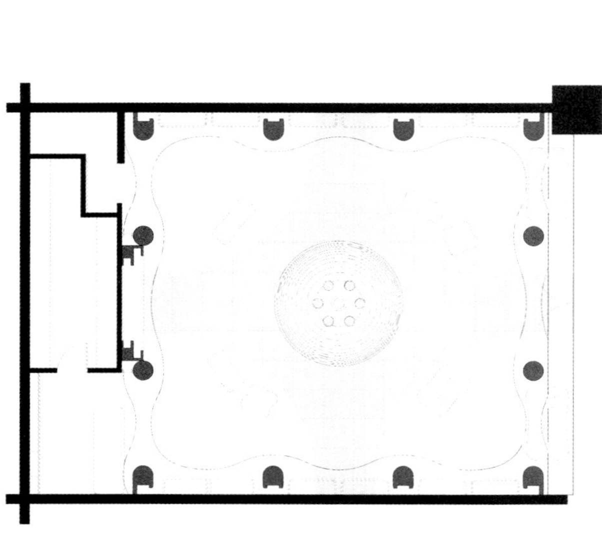

Fun Sports
Fun Sports
FUN SPORTS
Fun Sports FUN SPORTS
FUN SPORTS
Fun Sports Fun Sports

marco&mari

marco&mari

真维斯

设 计 师：迫 庆一郎、村田 知阳
设计公司：SAKO建筑设计工社
项目地点：中国无锡
项目面积：362平方米
摄　　影：沈忠海

庞大的物流支撑着现代都市的消费生活，而箱子应该可以视作是物流的象征。我们使用箱子来创作出能够刺激顾客购买欲的空间，以达到我们的创作目的。让我们想象，在100多平方米的白色空间里放入150个箱子，它们有的连接，有的分散，共同组合成一个三维立体空间。

客人进入店面时，店中内装陈设会引导客人体验跟一般的店中购物有所不同的购物乐趣。在这里，客人会感受到恍如进入到了入货仓库中卸货时的场景，能体会到在众多即将出货的货品中千辛万苦才淘到心头所好的欣喜和激动。

The enormous logistics support the consumption of the modern city; boxes can be seen as the symbols of logistics. Therefore, we use boxes to create the kind of space that stimulate the desire of the customers to purchase, which in turn achieves our target. Imagine that there are 150 boxes in a white space which covers more than 100 square meters, some of which are linked, some scattered, combining a three dimensional space.

Once the customers step into the store, the display will guide them to a fun, unique shopping experience. Different from the other stores, customers will feel like placing themselves in to storage where all the goods are unloaded, and feeling the thrill and happiness when they find what the love out of the enormous merchandise.

FULL OF
GREATEST

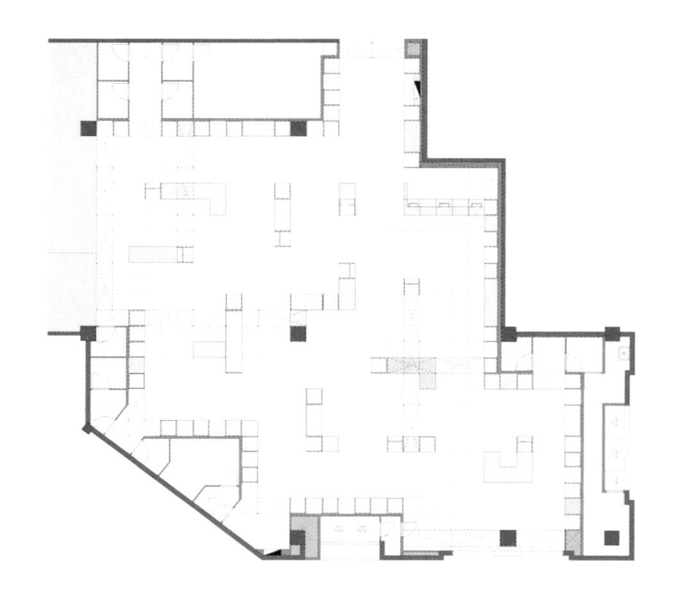

BMW Lifestyle

设 计 师：袁颖钊
建 筑 师：黑珍珠
设计公司：八六三建筑室内设计
项目地点：中国北京
建筑面积：500平方米
摄　　影：游英正

我们的设计概念，是由抽取我们最喜欢的最新宝马跑车M1、Z4和Z8那如雕塑般的外形开始，然后把这些车的造型巧妙地演绎成一个具有创意的零售空间。设计目的是创造一家在视觉上有很强动感、精致和不受时间洗礼的店，而它的灵感是来自宝马跑车的前腰形进气口，能让人从街上远处走近来看她。另外重要的是，同时创造一个很大能透视到店内的空间，且在店面创造有动感的橱窗展示，如那些天花的灯光凹槽造型和活动的平台。对于店的颜色设计考虑，就如为新车挑选颜色一样，我们试了很多方案。我们想有一种能让产品颜色展示它们最好一面的经典中性色调。同时，我们认为所用的色调，须与现有宝马车系所用的颜色互相调和，要能反映出客户的品牌身份。因此，我们选用了宝马跑车的颜色，并尽量在三种颜色中保留其中两种——白色和蓝色。那于墙面走上天花的蓝色造型斜条，让人立刻认出那是直接演绎了宝马M系跑车的商标。还有，所有用于店的油漆，是跟车厂使用的一样。

Our concept started by taking the sculptural qualities of our favorite recent BMW sports cars, the M1, Z8 & Z4 and to translate the forms of these cars subtly into an innovative retail environment.We aimed to create a store that had a strong visually dynamic, elegant and timeless shop front and were inspired by the kidney shaped front grill of BMW cars. We wanted to create a shop front that had the elegance and timelessness of this grill design yet create a design that would make one cross the street to see it closer. It was important to also create a large see through area into the store that would allow the clients VM (Visual Merchandizing) team a large staging area at the front of the store to create dynamic window displays, as such we have crated slots in the ceiling areas and movable platforms in the base of these window areas to aid with their displays.As with choosing the colour of a new car, we went through many options, we wanted a classic neutral colour that would allow the colour of products to be shown off to their best. We also felt the colours had to be in tune with current colours used on BMW models and also had to reflect the corporate identify of the client. As such we opted to follow the colours of the BMW motor sport division albeit keeping the colour palette to just two of the three colours reinweiss (white) and signalblau (blue). The diagonal blue stripe running up the wall is an instantly recognizable and direct interpretation of the inclined logo used on BMW M-series cars. All paint used within the store is the same as that used in the motor industry.

BMW Lifestyle

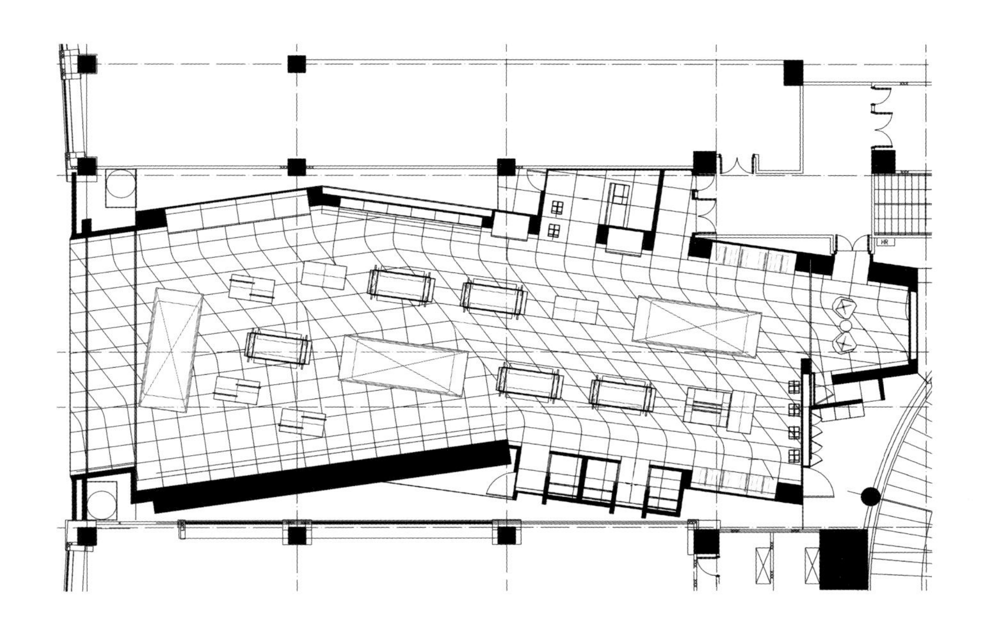

为了使顾客走进店内时能容易地看见整家店，货品能清楚地以多重层次来展示，我们创造了可让产品在高、中及低的高度展示的展架。我们的构思是让产品能展示于地上、于低处的展柜、于中间的挂架上或于高处的造型展墙。

所有的展台和附件配置，是由八六三根据它们的个别用途要求特别设计，包括展台、展柜、碳纤维挂架和造型展墙。矮展台设计成像时装表演的天桥，来展示行李箱和自行车。那倾斜的展柜是根据宝马M系的商标图案来设计，它可让细小的产品展示于玻璃展柜中。

Upon entering the store our aim was to allow customers to view from the front to the back of the store easily and to clearly display merchandize at multiple levels. Therefore we created furniture that allowed us to display products in a low, medium and high level format. Our idea was to have products displayed on the floor, at low level in show cases, at mid level hanging from rails or at high level displayed on the feature wall.

All display fixtures and fittings, i.e. podiums, show cases, carbon fibre hanging rails and the feature wall were custom designed by eightsixthree and are all specific to their individual requirement, low level podiums designed to look like fashion show catwalks are used to display luggage and bicycles, inclined (M-series) display cases are based on the BMW roundel logo and allow display of small items within a glass display case.

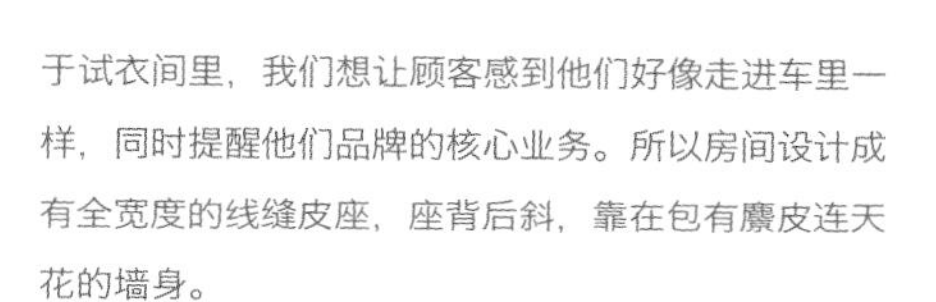

于试衣间里，我们想让顾客感到他们好像走进车里一样，同时提醒他们品牌的核心业务。所以房间设计成有全宽度的线缝皮座，座背后斜，靠在包有麋皮连天花的墙身。

这家店特设有贵宾房，于顾客的座位处有特别的影音设备，可让他们前往最近的特许销售店之前，查看和讨论特定车种的优点。

In the changing rooms we wanted customers to feel that they had just stepped into a car and be reminded of the core business of our client.

Therefore the rooms are designed with full width bench seats made from stitched leather set into a sloping back wall which is lined with alcantara which also runs across the ceiling as a head lining.

This particular store has a VIP room with a dedicated AV system where a client can sit, review and discuss the merits of a particular car before being directed to their nearest dealer.

EQ:IQ Flagship Store in Beijing

设 计 师：蔡明治
设计公司：蔡明治设计有限公司
公司网址：www.alexchoi.com.hk
项目地点：中国北京

EQ:IQ是中国主流女装品牌之一。他们的衣服风格简单但注重细节。该品牌的受众年龄广，从25岁到55岁。因此，我们为该店所做的设计采用"永恒的"的现代风格，然而细节和材质却很丰富。

旗舰店分为三个部分：鞋包专柜、年轻人服饰系列和主流服饰系列区域。该店选址是一个奇怪的"U"形，一个入口在购物商场内，另一个面向大街。主流系列安排在面向大街的那一侧，而直线形的区域就留给令人兴奋的年轻人服饰系列。鞋包专柜则安排在"更加"私密点的面向购物商场的一侧。

主流服饰系列区域最大的特点是豪华的马赛克图案的汉白玉墙。它的效果因为一个定制设计的"枝形吊灯"而加强了。枝形吊灯结合了含有坚固的白蜡木、亚克力棒、消光白色金属棒和黄铜棒的EQ:IQ材料，以及带有EQ:IQ树形标志的黑白地毯。简单的光滑金属白色旋转衣架用来陈列产品，与黏重质地的铅笔形成鲜明对比，产生一种微妙的"饱满"质感。

EQ:IQ is one of the leading Chinese apparel brands for women. Their clothes are simple in style but rich in details. It covers a wide age group of customers, from 25 to 55. As a result, the store design we create is in "Everlasting" modern style but rich in details & texture.

The flagship store is divided into 3 parts, the shoes & bags section, the young collection & the main collection. The site is an odd "U" shape with one entrance at the mall and another entrance facing the high street. The good point is we have a 2 storey high facade though we just occupy the ground floor. The main collection is on the side facing the high street, while the linear area is used for the exciting young line. The shoes & bags section is on the "more" intimate side facing the mall.

The main collection section is featured with luxurious white marble wall in mosaic pattern. It is reinforced by a custom design, "Chandelier", which is a combination of EQ:IQ materials with solid ash wood, acrylic rod, matt white metal rod and the brass rod, and the rug with black & white EQ:IQ tree icon. Simple and slick metal rolling racks in white are used for product display which contrasts with the heavy textured wall producing a subtle "Rich" quality.

EQ:IQ

107B

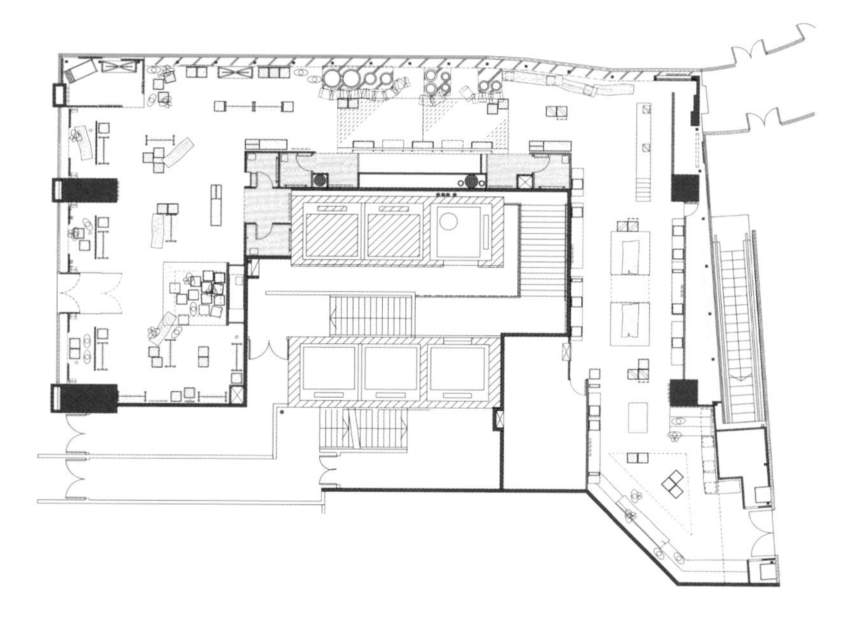

年轻人服饰系列区域最大的亮点就是管子构建起来的丛林。事实上，巨大的中断线是这个区域的特色，它们看起来像是从上方直接连通到商店里的管道。我们利用这一点将它和不锈钢制成的管状货架结合起来形成一个“丛林”。它与水平方向的松木板结构形成巨大反差，该结构是商品的家具也是用来遮蔽试衣区域的。

The young collection is highlighted by a jungle of tubular structures. In fact, the big trunks are the characteristics of the site which look like the ducts leading down from above to our site. We make use of it and integrate it with our stainless steel made tubular rack to form a "Jungle" feature. It works in great contrast with the horizontal pine wood plank structure which is a product fixture and screens the fitting area.

EQ:IQ

鞋包专柜设计成一个“电梯状”的结构，这一特点体现在产品夹具和店面陈列橱上，这个构想也是来自于这个空间的特点。这个空间位于扶手电梯的下方，白色的楼梯结构的设计不仅提供了动感的商品展示平台，也融合了原有的扶手电梯，使整个店面完整而有趣。利用EQ:IQ的树形标志定制设计的树形图案创造出一种包装纸的效果，统一了整个店面。

The shoes & bags section is designed with a "staircase-like" structure. This feature can be shown in the product fixtures & the shop front showcase. Again, this idea comes from the characteristics of this area. The area sits under a big flight of escalator. The custom design with white staircase structure at the shop front does not only provide a dynamic visual merchandising platform, but also blending with the landlord's escalator, making the whole shop front complete and interesting. A custom design with "tree" pattern by using EQ:IQ tree icon is used to create a wrapping paper effect to unite our shop front.

EQ:IQ Chain Store in Hangzhou

设 计 师：蔡明治
设计公司：蔡明治设计有限公司
项目地点：中国杭州
建筑面积：118平方米

在数年的成功经历之后，EQ:IQ的新面貌已经成功在中国各个不同地方推广开来。正如EQ:IQ的时装系列一样，设计理念也始终如一：风格简约却充满细节质感。
从入口进入这个漂亮的玻璃盒子，映入眼帘的是一盏订制的灯树。亚克力树脂与哑白色金属组合而成的树冠铺缀在树干上。树干本身原来是店内的结构柱，没有任何人工的修改，设计师将它利用起来设计成了EQ:IQ 的商标一树。简单的白色抛光金属旋转衣架，与质感丰富的墙壁形成鲜明对比，蕴酿了一种低调璀璨的气质。服饰置于陈列橱的透明架上，由垂直的黄铜管支撑着，与遮蔽试衣区域的水平松木结构相互交错形成反差。鞋子和包随意展示在摆放结构上，形成了一种拼凑效果的样式，这也创造出了更多的储存空间。这样的摆放结构更有利于自由组合，能够为连锁店增添新鲜感。

After the years of success, the new look of EQ: IQ has been successfully rolled out to different places in China, the design concept remains the same - much like EQ: IQ fashion collections - is minimalist in style but rich in detail and texture.
Accessible from the main entrance of a beautiful glass box where attentions are grabs immediately by a custom-made chandelier tree. The acrylic resin and matte white metal made cluster shade embellishes the trunk which was the structural column in the store, without any artificial touch up, designer takes advantage of it and it becomes the EQ:IQ branding icon - tree. Farther along and to one side, simple and polished metal rolling racks in white are used for product display which contrasts with the heavy textured wall producing a subtle "Rich" quality. Wears are showcased on transparent shelves supported by vertical brass tubing across from and contrasting with - a horizontal pine structure that screens off the fitting area. Shoes and bags are displayed on the random racking system which supports high density of storage and creates a patch-work-like pattern. The combinations of the racking system are infinite, so, rolling out the chain store may be boring, but a little taste of freshness will fix.

EQ:IQ
EQ:IQ
Sale

EQ:IQ
LOOK
WHAT'S
NEW

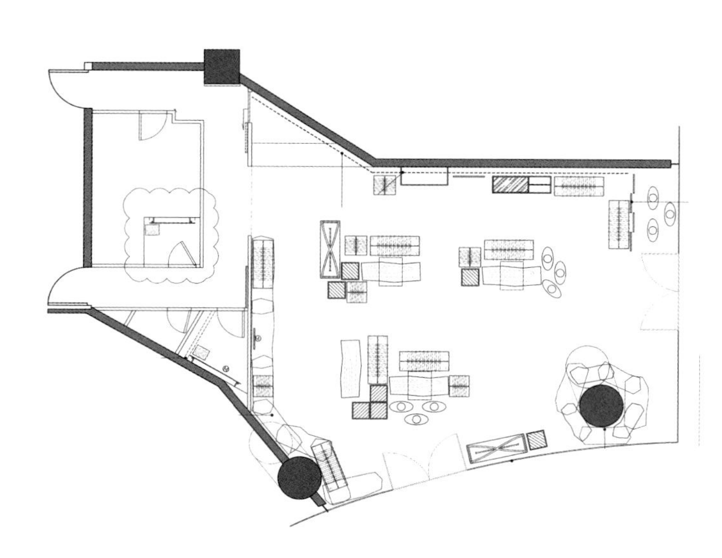

NEW ARRIVALS
EQ:IQ

EQ iQ

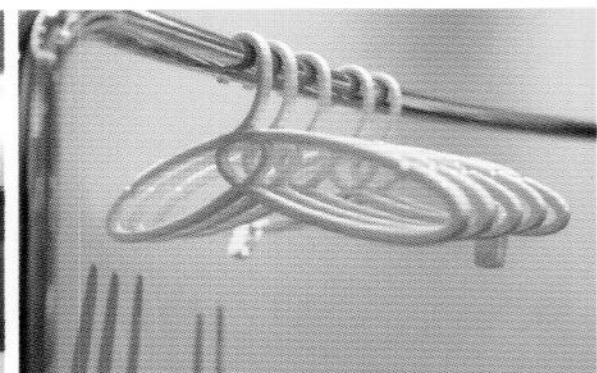

Alla Scala Flagship Store

设 计 师：欧比可建筑设计团队
设计公司：Oobiq Architects（欧比可建筑设计）
项目地点：中国深圳
建筑面积：175平方米
摄　　影：欧比可建筑设计

拥有创造另一成功品牌经验的客户对品牌定位十分清晰，Alla Scala的目标要求非常明确。作为品牌第一标识的标志，充满曲线的图案给客户很深刻的印象。这个标志就是我们展开这次创作的灵感泉源。

The requirements from Alla Scala were really clear, as they knew their positioning in the market, having the experience of another successful brand as their background. The logo, which should become something really well-known among customers, is really wavy with a lot of shapes. That was the main inspiration and the starting point in the approach to the design.

alla scala
8.8
8.8

由接到客户的委托开始，我们首先想到用色大胆鲜明的波普艺术，如Andy Warhol（安迪沃霍尔）的经典画作和50年代的家具。首个概念以Alla Scala的重点曲线铺满墙面。当代波普形象就此诞生，继而这些曲线化身成同心椭圆，让人联想到品牌标志中的A和L，椭圆的层次是由不同的物料和饰面所组成。这些椭圆并非单独出现在一个垂直面或水平面上。垂直面和水平面之间再看不出边界：椭圆装饰在垂直面和水平面中游走。椭圆装饰的主建材以温暖的木材配搭冰冷的不锈钢，形成强烈的反差。品牌主色挑选了年轻活力、出众抢眼，兼很少品牌在用的绿色。

When we started to think about this project, we had in our mind the strong colors of the pop art, like the painting by Andy Warhol, and the furniture which has been directly influenced by that cultural atmosphere. The first idea was to cover also the side walls of the shop with wavy shapes that would have been the main topic of the image of Alla Scala. A sort of contemporary Pop image was born. Then these shapes became concentric ellipses - another reminder to the A and L letters of the logo - where every part is made of a different material or a different finishing. These shapes are not only on vertical or horizontal surfaces. There is not anymore distinction between them: the decoration goes on the vertical and horizontal surfaces indistinctly, there are no more borders. The main materials we decide to use in the ellipses are the warm wood together with the cold steel, to create a contrast. Then of course the main color of the brand image, the green, a young dynamic color that is going to be outstanding as even if so attractive that not many companies are using it in their brand image.

alla scal

alla scala

概念内的所有物件设计均由曲线展开。如中岛陈列，就是按照小草所设计，有机图案不断贯穿整店。店内墙面以不同的绿色饰面创造出一个层次感丰富的动态空间。天然纯白的主货架和不锈钢次货架让本身质优的产品显得更亮眼。商场店的中厅将成为最吸引眼球的焦点，闪亮耀目的银面有机大造型在中厅创造出天然的洞。大造型外围设有挂产品的空间，充满神秘感的洞内同时可放置多于一个模特或产品组合。整个概念设计的重点在于四周的弧形曲线创造出客人被拥抱的舒服感觉。所有细节均沿用同一个概念，为这个品牌建立一个概念完整、非凡出众及独一无二的形象。

All the items in the concept are designed according to the idea of the wavy shapes. All the free standing displays, for example, are designed according to a L-shape that create an organic design all around the shop. The walls of the shop are treated with green surfaces of different finishing that can create a very dynamic and constantly changing space. Main shelves will be in a neutral white, so that products will be even more outstanding, or in stainless steel, with smaller sections and size. The middle hall in the bigger showroom will be used as an eye-catching focus area. An organic sculpture, covered with a shiny silver color, will be designed as a natural cave in the middle of the shop. Outside it will be also a space for hanging products, while in the middle there will be one or more manikins, as a precious island, hidden from the main part of the customers. The basic idea of the concept was to create an environment where customers can feel comfortable, where the round shapes will give an idea of warmly embracing people. All the details are designed according to the same concept, in order to create a brand which is outstanding and unique on the market.

ESS
AL
Anna Lochy
8.8

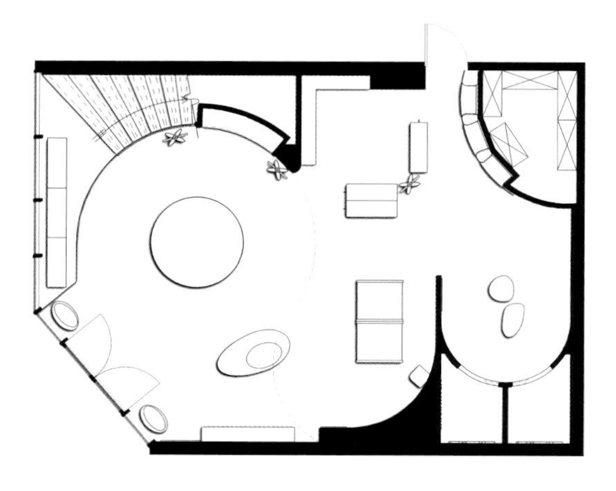

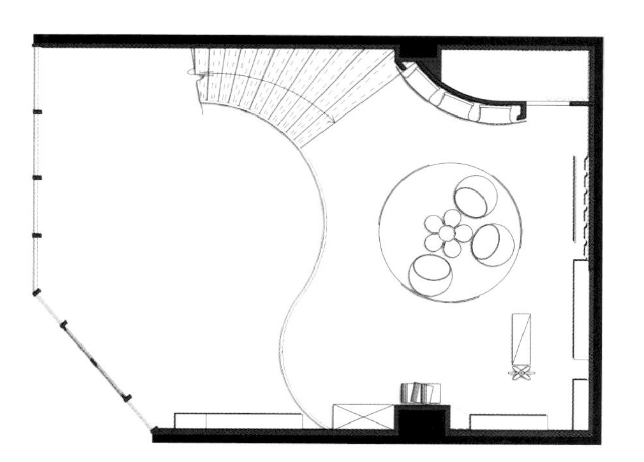

Giammabruns Flagship Store

设 计 师：欧比可建筑设计团队
设计公司：Oobiq Architects（欧比可建筑设计）
项目地点：中国重庆
建筑面积：500平方米
摄　　影：高寒

Giammabruns意大利精品屋是一家新成立的公司，主要在中国内地销售一些在欧洲非常受欢迎的意大利品牌服饰。
在设计师与业主面谈分享了他们对这个项目的基本想法之后，主要目标就非常明确了：为这些商店打造一个带有欧式的复古情怀的现代形象。
然而，整体环境又必须尽量从简，因为店里的主角是商品而不是商店本身。
具体说来，店里的时尚商品都有很强的个性，因此设计师就想到了打造一个最简单的"盒子"来展示这些商品，由此形成两者之间的强烈对比：简洁的店铺设计，丰富的服装样式。
重庆的这家商店是目前Giammabruns 意大利精品屋最大的一家，位于市内最成功的商业区之一，建筑面积约500平方米。品牌的理念总是在正面的墙上妆点欧式风格：富有光泽的木质表面，装饰着法式风格的细木护壁板，在灯光的映照下，墙面和木质表面之间的空间似有若无，飘浮在空中。

Giammabruns Italian Boutique is a new settled company that will distribute in Mainland China trendy Italian brands that are really popular in Europe.
Once Oobiq Architects directors "Giambattista Burdo and Samuele Martelli" met the ownership and shared the basic ideas of this project, the main target was quite clear: to create a modern image for these shops, with evident memories of European style.
But the whole environment had to be minimal as the main character in the shops should always be the goods, not the shops themselves.
In particular, the fashion items sold in these shops have a very strong identity, so the basic idea of Giambattista Burdo and Samuele Martelli was to create a minimal box where to display them, in order to create a contrast between the richness of the clothes and the minimal design of the shop.
This Chongqing shop is the biggest of Giammabruns Italian Boutique till now.
It is about 500 squared meters, inside one of the most successful commercial spaces in town.
The concept for the brand plans always a wall on the fa?ade that remind of Europe: a glossy white wooden surface with decoration that reminds of French boiserie is detached from the walls, as floating on the air, thanks to the lights around that make the space between wall and wooden surfaces as disappearing.

GiammaBruns

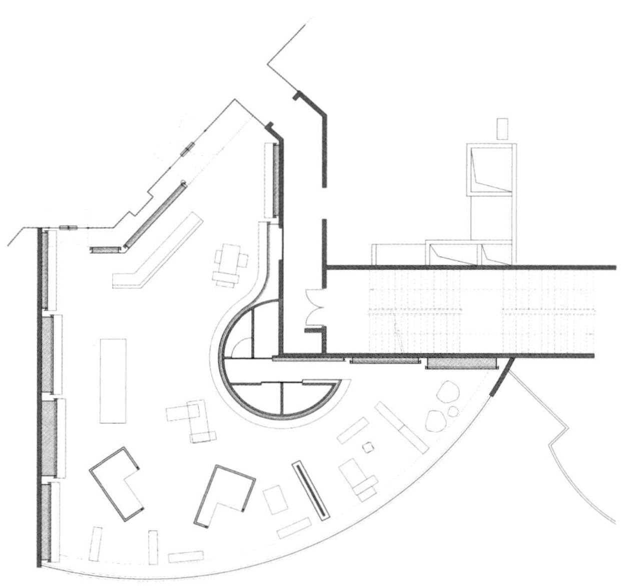

根据品牌需求而设的大小不一的独立式展柜更是别有一番天地。形状依然小而简，好像一个环，表面跟墙面的感觉很接近，而里面则别有洞天。
铜质表面应用在收银台的底部及吊灯的内部。一切都是设计公司专门为这个品牌量身定制的。因此里面所有的东西都表达了相同的感受，起到相同的装饰效果。
地板丰富而简约，大理石地板上铺着珍贵的浅棕色地毯，与后面的墙面相呼应。棕色调的运用使得这个商业空间充满了温馨温暖的气氛。
因此，所有的理念都基于对比：冰冷的白色和温暖的棕色，简约设计和法式壁板，简约的白色装饰面和奢华的铜质表面，光滑的大理石和富有质感的地毯。

The free standing displays, of different sizes to fit the requirements of the brand, introduce something more. The shape is still very minimal, a sort of loop, where the outside surface is very close to the feeling of the walls, while inside there is a sort of surprise. A richer material, more precious and so hidden is covering all inside the free standing display.
This copper surface is also the material for the bottom of the cashier desk and also the inside of the hanged lights. Everything is custom-made for this concept, designed by Oobiq Architects especially for this brand, so all the objects present the same feeling and same finishing.
The floors are really rich and minimal as the entire environment. Marbles are spaced out with precious carpets of light brown tones, the same on the back of the walls. The use of these tines of brown are used to warm up the atmosphere of the commercial spaces.
So all the concept is based on contrasts: the cold white and the warm brown, the minimal design and the French boiserie, the simple white finishing and the rich copper metal surfaces, the smooth of marbles and the textures of carpets.

Song of Song

设 计 师：欧比可建筑设计团队
设计公司：Oobiq Architects（欧比可建筑设计）
项目地点：中国深圳
建筑面积：140平方米
摄　　影：欧比可建筑设计

Song of Song 是深圳影儿集团精心打造的奢侈时尚品牌，该企业在中国时装市场具有领导地位。
影儿集团指定欧比可建筑设计为Song of Song进行改造以提高品牌地位。第一代店面形象欠缺冲击力，基本元素为白色木制墙体装饰及光洁明亮的大理石地板，配以简单的不锈钢衣架。因此，欧比可建筑设计的首要任务是构思如何以一个新颖有趣的方式运用原有建材以保留品牌印象。
经过欧比可设计团队努力创作，一个新设计诞生。白色木板搭配体现女性温婉的深灰色天鹅绒，重叠的矩形图案成为店的重点装饰，亦同时变成这个品牌的新标志。这个图案主要出现在橱窗背景和店的四周。展示柜的结构简洁，选材精致，这有助突出柜内陈列的珍贵首饰。衣架保留不锈钢的材质，但它被设计成一条缎带绑上覆盖着灰黑色天鹅绒的壁柜。中岛挂衣架也是同系列的设计。地板材料选择明亮雅致的棕色大理石。新概念与材料运用完美，呈现品牌奢华优雅的形象。

Song of Song is the luxury fashion label created by Shenzhen Yinger Group, one of the most important enterprises in Chinese fashion market.
Yinger Group appointed Oobiq Architects to upgrade the retail identity of Song of Song. The original store image was not too strong, made basically of white wooden panels on the wall, shiny bright marbles on the floor and simple stainless steel hangers. So the main target for Oobiq Architects was to think how to use the materials which was representing the brand but in a new and more interesting way.
After the creative process by Oobiq Architects' talented design team, a new design born. The white wooden panels come together with new parts in soft and feminine dark grey velvet, while a pattern of overlapping rectangles becomes the main decoration of the shop, a sort of new logo for the company. This decoration also appears on the background in all shop windows and some areas in the shop. The showcases present in a minimal structure where the only precious detail is left to the material. This helps to make more evident on the preciousness of the accessories. Hangers are still in stainless steel, but it's designed as a tape embracing the niches in the wall, which are covered with dark gray velvet. The stainless steel is applied on the free standing hangers also. A travertine marble in bright with elegant tones of brown is the material chosen for the floor. The new concept and the use of materials perfectly show the brand identity, luxury and elegance.

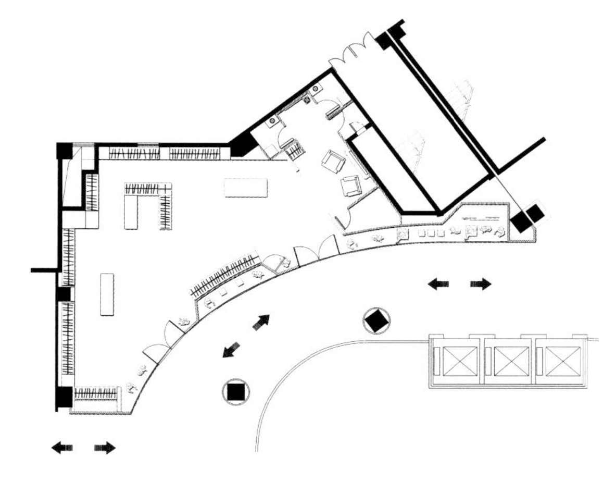

Giada

设 计 师：丁晓斌
项目地点：中国深圳
建筑面积：186平方米

设计师倾力打造了Giada亚太区的多个专卖店，成功演绎Giada的奢华细致。
淡淡的优雅气氛，在暖色调集结的瞬间，显露出Giada意境之美。有切割的空间、有节奏的摆放、精确的光影设计延伸出立体的视觉空间，展现着Giada的优雅之美。宽敞明亮的空间感打造奢华的购物环境，精致的橱窗给人以灵动美丽的印象。欧洲浅栗色胡桃木配上白色皮质的高雅点缀，打造Giada专卖店全新的高端奢华新概念。

The designer made all efforts to design many Giada exclusive shops in APAC, successfully interpreting luxury and meticulousness of Giada.
The subtle elegancy atmosphere reveals the beauty of artistic conception of Giada at the moment when warm tones concentrate. The space with incision, rhythmic arrangement and precise shadow design roll out three-dimensional visual space, revealing the elegant beauty of Giada. Spacious and bright sense of space created luxurious shopping environment. Exquisite showcases give people a vivid and beautiful impression. The European light maroon walnut dotted with elegant white leather creates brand-new high-class luxury concept of Giada exclusive shops.

GIADA
免税商場
KALTENDIN
KALTENDIN

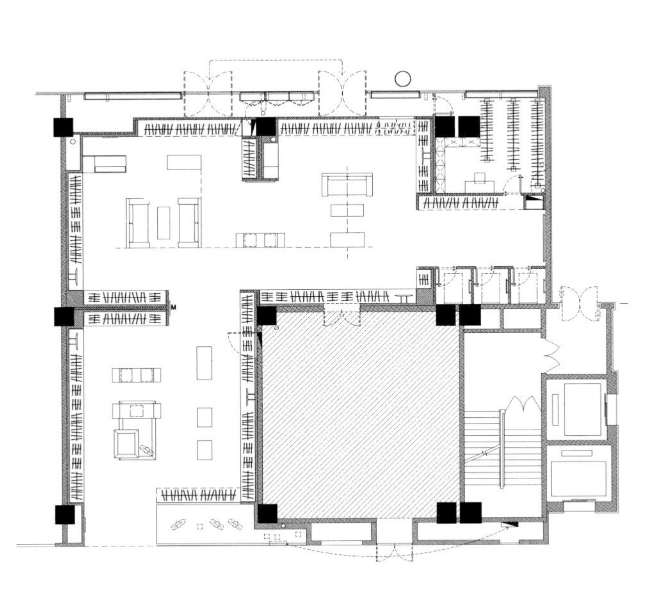

GIADA

Trend Platter

设 计 师：Lam Wai Ming（林伟明）
设计公司：Design Systems Ltd.
公司网址：www.designsystems.com.hk
项目地点：中国上海
建筑面积：350平方米

这个项目是一个年轻潮流品牌位于中国上海的商店，面积约350平方米。

现今的商店设计，不仅仅只是由设计师单方面从空间或功能作出理解和进行设计。在设计过程中，客户和设计师相互之间产生的化学作用更能迸发出特别的火花。具体来说，客户让设计师了解他们的顾客群、竞争者、产品特色及市场定位。然后，设计师会根据他们的特点，为他们创造一个有意义、美观及持久耐用的设计。

为了实践这个理念，我们与客户同时认为应找出一个特别的做法来演绎品牌的“多样化”和“混搭”特点。整个室内空间设计成一个潮流拼盘，多种多样的设计风格，包括古典的、现代的和怀旧的风格。

在不同风格的背景衬托下，即使品牌内不同系列的服饰风格相近，也能产生不同而又丰富的视觉效果。换句话说，同一件衣服在不同的背景下也可营造出不同的感觉，我们认为这也是一种突出此品牌特性的创新手法。

Located in Shanghai, China, this is a boutique for a young fashion brand with area about 350 square metres.

Nowadays, retail design is more than spatial or functional design which is just unidirectional work from the designers. The interactions between clients and designers will strike more gorgeous sparks. Namely, clients help designers to understand their customers, competitors, originations, and marketplace; in turn, designers try to emphasize their distinctiveness by creating meaningful, aesthetically pleasing, and enduring design solutions.

To implement this notion, design inspiration of this project was drawn from the brand characters "Varieties" and "Mix and Match". As if a trend platter, visual features within the space were created with a juxtaposition of a variety of trendy design styles including classical, modern, vintage and minimal.

By matching the garments with different style of backgrounds, visual richness can be created even there is a certain degree of similarity among various collections of garment. We see this as an alterative and creative way to highlighting the trait of the brand.

+−×÷
ODF

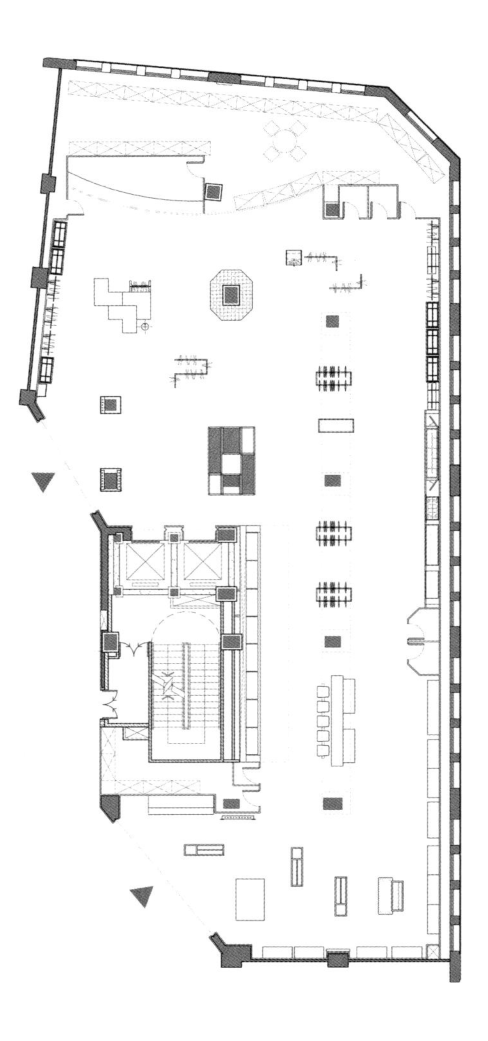

红贝缇

设 计 师：梁兆新
设计公司：新贤维思设计顾问有限公司

店面内装设计的好坏，评判的标准在于空间规划的动线是否合理，设计出来的效果是否能勾起目标客户群体的购买欲望。本案的设计师在面对这样一个方正的店面空间时，选择从灯光入手，对中岛柜的货架展示区采取集中照明。用这样一种高低错落的吊灯群设计首先营造出一种阵势，容易将入店客户的视觉焦点吸引至此。其次，这样堆积的光线照度打亮了引导消费的主打服饰。将服装店的销售目标准确锁定，让目标客户一目了然。空间规划上，设计师利用商业店面高挑的空间，做出一个空间错层，丰富了空间的立体效应；在空间功能规划上将一层店面设计为服饰的集中展示引导消费，利用曲折的楼梯将选中衣裳的目标客户引导至二层进行体验式消费，并巧妙地利用了沿梯走道进行目标指引。这样，目标客户可以在特意做出的二层空间里试衣、休息，持续选衣。一系列的消费引导后，目标客户群在心满意足的服饰体验后于楼梯口处买单完结消费行为。纵观整体空间，设计师设计手法熟练，工作流程规划合理，充分体现了商业设计的促销要素，达到了业主需要的设计产品。

The criteria for judging interior design of a shop is whether the moving lines of space planning are reasonable and whether the design effect can appeal to target customer groups to buy. In the face of an upright and foursquare store space in this project, the designer chose to begin with the lighting. The concentrated illumination was used in the shelves display area of middle counter. First of all, the design method that the group of droplights was allocated in either high or low positions created a momentum, easily attracting the visual focus of the customers who enter the store. Secondly, the cumulate light illumination lightens the leading clothes that guide consumption, accurately locking the sales target of the fashion store and making the target customers clear at a glance. As for the space planning, the designer made a space split level taking advantage of the high space of commercial store, enriching the stereoscopic effect of the space; and for the planning of space functions, the first floor storefront was designed for the intensive display of clothes to guide consumption. The target customers who have chosen certain clothes will be leaded to the second floor by a twisting staircase for experimental consumption. The corridor along the staircase was skillfully used for target steering. In this way, the target customers can try on clothes, have a break and continue to choose clothes in the specially made space on the second floor. After a series of consumption guide, the target customers pay the bill at the foot of staircase to end up the consuming behavior after a satisfactory clothing experience. Making a general survey of the space, the designer has a sleight of hand and his workflow was well arranged, fully reflecting sales promotion factor of commercial design and making a design product that meets the need of the owner.

HON.B

HON.B

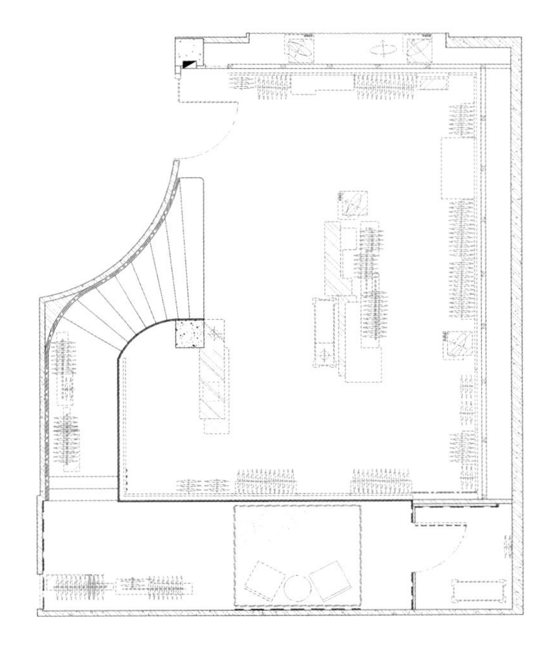

米柯拉

设 计 师：梁兆新
设计公司：新贤维思设计顾问有限公司

商业空间的设计需要设计师能提供给目标客户一个良好的购物环境和一个顺畅的购物流程。面对门面不大且曲折纵深的空间，设计师在规划上大胆地在门面部分的左侧开辟出一个具有十足展示效果的时装伸展台。在这个时装舞台上，设计师将时装设计师的设计理念及将要在这一季推介的作品展示，以引起购买客户的兴趣。紧接着设计师在顶部的处理上采用阵列的灯具形式进行指引，利用导购中岛台引导客户入店选购。在经过一番曲折后豁然开朗，来到宽大的正店中心。在此间里，设计师没做过多的设计堆砌来展示销售服饰，而是采取一种大开大合的方式，用一种宽松的氛围来引导客户独一性的尊贵享受。此间里的服饰展示高低错落，稀疏摆放，特意营造出一种少而精的购物氛围来提升销售产品的档次。整个设计简约、时尚、高档，强调出了业主所需要的精品销售的设计理念，达到了商业销售所需要的设计效果。

The design of commercial space requires the designer to provide target customer with good shopping environment and smooth shopping process. Coping with the winding and deep shop with small façade, the designer audaciously created a fashion runway which does not bring about profit but has wonderful display effect at the left of the shop façade when planning. On the fashion runway, the designer displayed the design concept of fashion designer and the works marketed in this season to attract the customers to buy. With that, when disposing the roof, the designer showed by means of the display of lamps and lanterns, guiding the customers to enter the shop to purchase by shopping guide middle counter. After the winding, the space suddenly becomes extensive, leading to the spacious shop center. In this space, the designer did not pile too many designs to display and sell clothes, but guided customers to exclusive and distinguished enjoyment in an overall manner with loose atmosphere. In this shop, the clothes are displayed in both high and low position sparsely to create a shopping atmosphere of smaller quantity and better quality and upgrade the goods sold. The overall design is simple, fashionable and high-class, highlighting the design concept of quality goods sales that the owner needs and achieving the design effect of commercial sales.

MIK&LA

MIK&LA

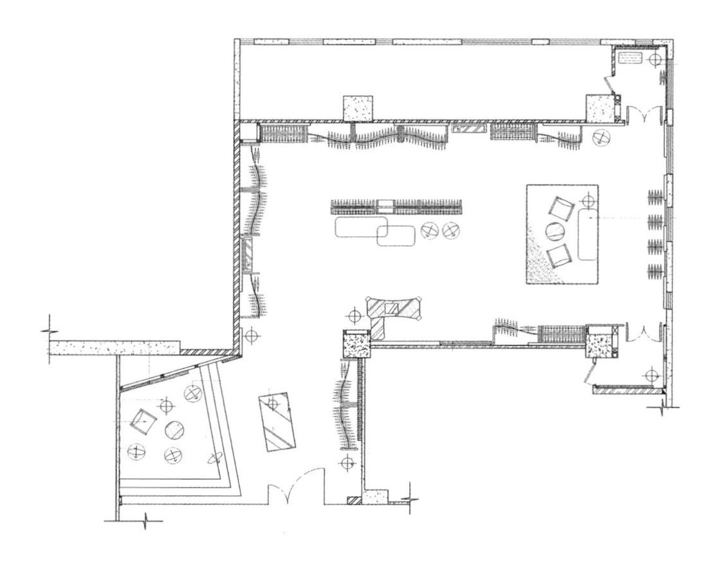

MIK&LA

女友时装店

设 计 师：郑少文
设计公司：汕头市博一组设计有限公司
项目地点：中国汕头
建筑面积：130平方米

水养育生命如母亲，温柔如少妇，清澈如少女。所以说：女人若水。这是对女人最好最美的形容。
女友时装店是经营30岁至40多岁的女人服饰。这个年龄段是女人最具女人味的阶段，所以女友时装店用“水”作主题是最合适的。柔顺的曲线如水的波澜，是美的律动；圆形的层架如水泡向上飘升，是未泯灭的梦想；如水滴一般的吊灯错落有致，平添几分浪漫；纯净的白色调是成熟的淡定。

Water, is as great as mother, as tender as woman and as pure as maid. Therefore, woman is water. This is the best term to describe the women.
Girlfriend boutique targets at women age from 30s to 40s. Women in this age group are considered as the most feminine. Therefore, water fits the most the theme of Girlfriend boutique. Submissive curves and serous waves are the rhythm of beauty; round shelves fluttering up like water bubbles are ever-lasting dreams; water-drop-like droplights are well-spaced, enhancing the atmosphere with romance; pure white color symbolizes the calmness of maturity.

GIRLFRIEND

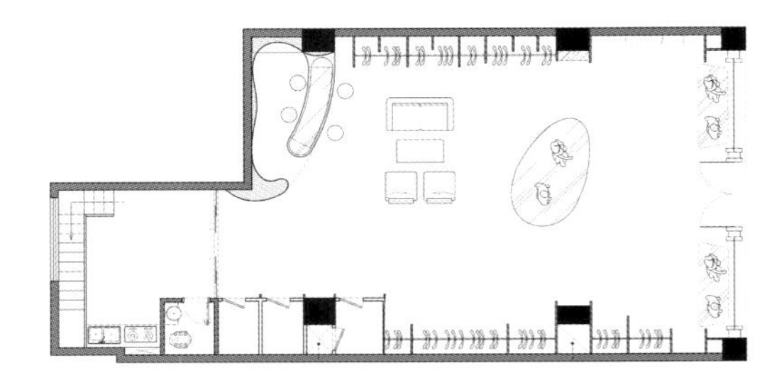

GIRLFRIEND

TP国际名品旗舰店

设 计 师：李云山
方案审定：叶斌
设计公司：福建国广一叶建筑装饰设计工程有限公司
项目地点：中国福州
项目面积：300平方米

TP国际名品旗舰店位于城市中心繁华地段，设计师利用了原有结构楼层高的特点，在保证了前厅大气、高挑的气势后，将店铺的后半区使用钢结构隔层，划分出了一层男装区域及二层女装区域，既丰富了空间美感，又增加了空间利用率。二层的层高偏低，所以设计师在吊顶的处理上，以暴露式设计为主，保留了原有结构梁的构造，巧妙地做一些黑镜造型吊顶，既美化了空间，又保证了高度。

本案的另一特点：大量利用了镜面反射的魅力，不管是造型门的银镜饰面，背景墙的大量黑镜应用，以及二层过道的落地镜，还有细节处的镜面不锈钢装饰，都充分体现了空间延伸感以及时装卖场特有的时尚气息。

在色彩的应用上，本案以黑、白、灰为基调，如：大面积的黑镜，米白色的法国木纹石搭灰色玻化砖，以及大气的白色软膜天花，无处不在地为多样化的时装创造出一个迷人的展示空间。

TP international brands flagship store is located in the busy section of downtown area. The designer made use of the characteristic of the high floor in original structure to ensure the grandness and height of the lobby. The back area is separated from the front by steel structured interlayer, marking out the men's wear section on first floor and the women's on the second floor, enriching the space beauty as well as increasing space utilization rate. Result from that, the height of the second floor is a little low. The designer execute exposure style mainly in dealing with ceiling, reserved the original structure of beams and made some dark mirror suspended ceiling tactfully which beautify the space as well as ensure the height.

The other characteristic of the project is the massive use of mirror reflection glamour. The silver mirrored decoration face of modeling door, massive use of black mirror on background wall and floor mirror in corridor on second floor as well as mirrored stainless steel decoration in detailed place have fully reflected space outspread feel and unique fashionable atmosphere in fashion store.

In the color application, the project use black white and grey as color motif. For example: large scale of black mirrors, beige serpeggiante matching grey polished tile, and grand white mantle ceiling, here and there, creating a charming display space for the various fashion.

TP 国际名品

D&G
PRADA
PAUL & JOE
FERRE
PATRIZIA PEPE

FERRE
PATRIZIA PEPE

D&G
DSQUARED²
FERRE
PATRIZIA PEPE
FIRENZE
BIKKEMBERGS
DOLCE&GABBANA

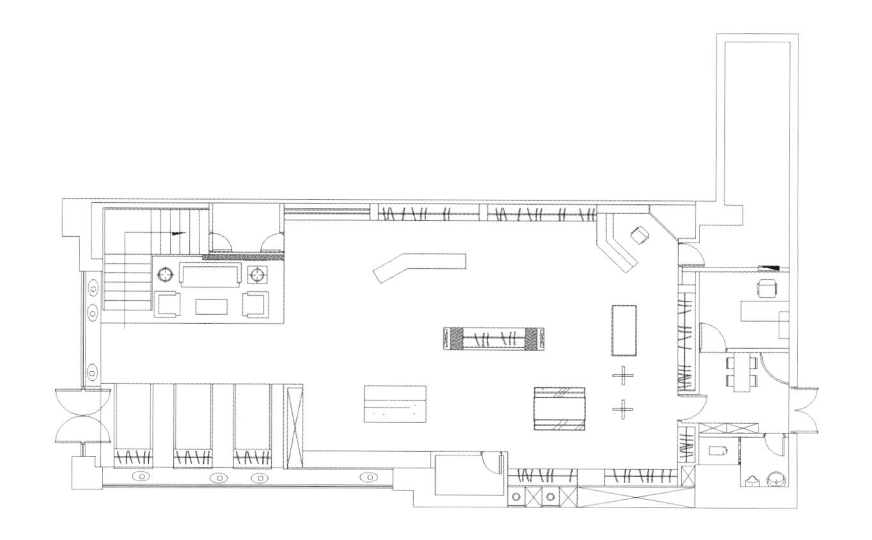

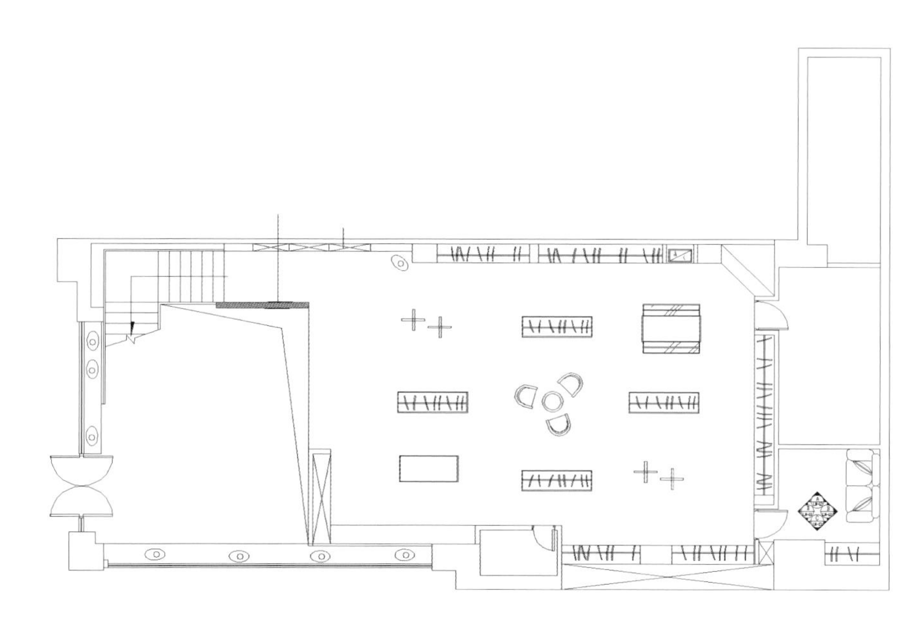

图书在版编目（CIP）数据

国际顶级服饰名店 / 深圳市创扬文化传播有限公司编 . —福州：福建科学技术出版社，2011.7
ISBN 978-7-5335-3829-3

Ⅰ . ①国… Ⅱ . ①深… Ⅲ . ①服饰 – 商店 – 室内装饰设计 – 作品集 – 世界 Ⅳ . ① TU247.2

中国版本图书馆 CIP 数据核字（2011）第 061969 号

书　　名　国际顶级服饰名店
编　　者　深圳市创扬文化传播有限公司
出版发行　海峡出版发行集团
　　　　　福建科学技术出版社
社　　址　福州市东水路 76 号（邮编 350001）
网　　址　www.fjstp.com
经　　销　福建新华发行（集团）有限责任公司
印　　刷　恒美印务（广州）有限公司
开　　本　635 毫米 ×965 毫米　1/8
印　　张　40
图　　文　320 码
版　　次　2011 年 7 月第 1 版
印　　次　2011 年 7 月第 1 次印刷
书　　号　ISBN　978-7-5335-3829-3
定　　价　298.00 元